# Our Biometric Future

CRITICAL CULTURAL COMMUNICATION

*General Editors:* Sarah Banet-Weiser and Kent A. Ono

*Dangerous Curves:*
*Latinas in the Media*
Isabel Molina-Guzmán

*The Net Effect:*
*Technology, Romanticism, Capitalism*
Thomas Streeter

*Our Biometric Future:*
*Facial Recognition Technology and the Culture of Surveillance*
Kelly A. Gates

# Our Biometric Future

*Facial Recognition Technology and the Culture of Surveillance*

Kelly A. Gates

NEW YORK UNIVERSITY PRESS
*New York and London*

NEW YORK UNIVERSITY PRESS
New York and London
www.nyupress.org



References to Internet websites (URLs) were accurate at the time of writing. Neither the author nor New York University Press is responsible for URLs that may have expired or changed since the manuscript was prepared.

Library of Congress Cataloging-in-Publication Data

Gates, Kelly.
Our biometric future : facial recognition technology and
the culture of surveillance / Kelly A. Gates.
p. ; cm. — (Critical cultural communication)
Includes bibliographical references and index.
ISBN 978-0-8147-3209-0 (cl : alk. paper) — ISBN 978-0-8147-3210-6
(pb : alk. paper) — ISBN 978-0-8147-3279-3 (e-book)
1. Biometric identification. 2. Face—Identification. I. Title.
II. Series: Critical cultural communication. [DNLM: 1. Biometric Identification.
2. Technology. 3. Facial Expression. 4. Security Measures—ethics.
5. Social Environment. 6. Terrorism—prevention & control. TK 7882.S65]
TK7882.B56G38 2011
363.2'32—dc22 2010039825

New York University Press books are printed on acid-free paper,
and their binding materials are chosen for strength and durability.
We strive to use environmentally responsible suppliers and materials
to the greatest extent possible in publishing our books.

Manufactured in the United States of America
c 10 9 8 7 6 5 4 3 2 1
p 10 9 8 7 6 5 4 3 2 1

*For my mother, Thelma Eileen Gates*

# Contents

# Abbreviations

| | |
|---|---|
| AFEA | Automated Facial Expression Analysis |
| *AMW* | *America's Most Wanted* |
| ARL | Army Research Laboratory |
| BCC | Border Crossing Card |
| CBIR | Content-based image retrieval |
| CMU | Carnegie Mellon University |
| CRM | Customer relationship management |
| DARPA | Defense Advanced Research Projects Agency |
| DMV | Department of Motor Vehicles |
| DVR | Digital video recorder |
| EFT | Electronic funds transfer |
| EMFACS | Emotion Facial Action Coding System |
| FACS | Facial Action Coding System |
| FACSAID | Facial Action Coding System Affect Interpretation Dictionary |
| FERET | Face Recognition Technology program |
| fMRI | Functional magnetic resonance imaging |
| FRVT | Facial Recognition Vendor Test |
| HCI | Human-computer interface |
| ICT | Information and communications technology |
| INS | Immigration and Naturalization Service |
| INSPASS | Immigration and Naturalization Service Passenger Accelerated Service System |
| IPTO | Information Processing Techniques Office |
| NEC | Nippon Electric Company |
| NIST | National Institute for Standards and Technology |
| SPOT | Screening Passengers by Observational Techniques |
| TIA | Total Information Awareness |

| | |
|---|---|
| TPD | Tampa Police Department |
| TSC | Terrorist Screening Center |
| UCSD | University of California, San Diego |
| US-VISIT | United States Visitor and Immigrant Status Indicator Technology |
| YCDC | Ybor City Development Corporation |

# Acknowledgments

This book began as a research project in the Institute of Communications Research at the University of Illinois, Urbana-Champaign, and I wish to thank all my friends and colleagues who made my experience there so personally rewarding and intellectually transformative. I am also grateful to my colleagues in the Media Studies Department at Queens College, and in the Department of Communication and the Science Studies Program at University of California, San Diego.

The people who have had the most influence on the pages printed here are James Hay, Cameron McCarthy, Toby Miller, John Nerone, Jonathan Sterne, Paula Treichler, Dan Schiller, and Angharad Valdivia. They have been extraordinary intellectual guides, and their scholarship has served as models for my own. Conversations with Craig Robertson have likewise made a deep imprint on this book, and Val Hartouni read the complete manuscript and provided valuable feedback at a critical stage. I also owe a major debt of gratitude to Marie Leger, for making the kind of immeasurable contribution that only one's closest, most trusted friend can make.

I consider myself very lucky to have worked with such amazing people at NYU Press: Eric Zinner, Ciara McLaughlin, and Despina Papazoglou Gimbel, as well as the editors for the Critical Cultural Communication series, Sarah Banet-Weiser and Kent Ono. Others to whom I offer special thanks are Mark Andrejevic, Ted Bailey, Boatema Boateng, Jack Bratich, Barry Brown, Lisa Cartwright, Tamara Chaplin, Cliff Christians, Sue Collins, Jennifer Dross, Rachel Dubrofsky, Grace Giorgio, Brian Goldfarb, Nitin Govil, Dan Hallin, Larissa Heinrich, Sonja Hillman, Robert Horwitz, Sharon Kehnemui, Marina Levina, David Lyon, Shoshana Magnet, Richard Maxwell, Shawn McLaucic, Dan McGee, Jen Mercieca, David Monje, Laurie Oullette, Jeremy Packer, Lisa Parks, Victor Pickard, Joy Pierce, Carrie Rentscher, Gil Rodman, Max Rodriguez, Francine Scoboria, Ben Scott, Sherri Sebastian, Kumi Silva, Rob Sloane, Gretchen Soderlund, Joseph Turow, Andrew Whitworth-Smith, Sharon Zupkin, and my writing group for the CUNY Faculty Fellowship Publications Program.

I also would like to thank my family, who gave me unending support during the years I spent working on this project: my mother, Thelma Eileen Gates, to whom I dedicate this book; Todd Gates, Andrea Gates, Rege Hovan, Traci Lersch, Alex, Austin, Shawn, Haley, and Aaron; as well as my dad, Stan Bernadowski, and my other family, Judith Tener and David Lewis. In addition, there is no way I could have completed this book without the volumes of encouragement and insight I received from David Philip Tener. I cannot adequately express how much I owe David for the countless hours of unpaid mental and emotional labor he provided in support of this book and its author. And finally, special thanks to Greg Barrett, who made the very end of this project an education and experience unto itself.

A still image taken from surveillance video at the Portland, Maine, airport on the morning of September 11, 2001, appears to show two of the alleged hijackers, Mohammad Atta and Abdulaziz Alomari, passing through airport security. Courtesy Getty Images. Reprinted with permission.

# Introduction

## *Experimenting with the Face*

> Of all the dramatic images to emerge in the hours and days following the September 11 attacks, one of the most haunting was a frame from a surveillance-camera video capturing the face of suspected hijacker Mohamed Atta as he passed through an airport metal detector in Portland, ME. Even more chilling to many security experts is the fact that, had the right technology been in place, an image like that might have helped avert the attacks. According to experts, face recognition technology that's already commercially available could have instantly checked the image against photos of suspected terrorists on file with the FBI and other authorities. If a match had been made, the system could have sounded the alarm before the suspect boarded his flight.
>
> —Alexandra Stikeman, "Recognizing the Enemy," *Technology Review*, December 2001

The September 11 terrorist attacks generated an enormous flood of imagery, and among the deluge was a grainy shot of two of the alleged attackers taken early that morning at a security checkpoint in the Portland, Maine, airport. The recorded video image, which appears to show Mohammad Atta and Abdulaziz Alomari passing through airport security, is a familiar part of 9/11 iconography. Although difficult to discern the men's faces in the image, it is virtually impossible to reference it without also invoking the claim that facial recognition technology could have identified the men as wanted terrorist suspects. Already existing commercially available systems, according to this regretful yet strangely hopeful assertion, "could have instantly checked the image against photos of suspected terrorists" and alerted airport security.[1]

The suggestion that an automated facial recognition system may have helped avert the September 11 terrorist attacks was perhaps the most ambitious claim circulating about biometric identification technologies in the aftermath of the catastrophe. The precise origin of the claim is hard to identify; it seemed to spring forth simultaneously from multiple sources. If it first came from someone in the security industry, it was quickly embraced and repeated by other public figures who felt sure it was true. This hopeful, regretful possibility was the basis for hearings held on Capitol Hill following September 11. On November 14, 2001, a Senate subcommittee on Technology, Terrorism, and Government Information held a hearing on "Biometric Identifiers and the Modern Face of Terror: New Technologies in the Global War on Terrorism." In her opening remarks, Senator Dianne Feinstein (D-CA) asked, "How could a large group of coordinated terrorists operate for more than a year in the United States without being detected, and then get on four different airliners in a single morning without being stopped?" The answer, she noted, "is that we could not identify them." Voicing again the assertion that had become part of the repertoire of public responses to the 9/11 events, she asserted, "In the case of at least two of the hijackers, authorities had pictures of them as suspects prior to the attack, and airport cameras actually photographed them. But because these cameras didn't use facial biometric systems, security was not alerted and the hijackers remained free to carry out their bloody plans."[2]

The idea that the events of 9/11 could have been prevented with the sophisticated technological products of modernity was laden with what the cinema studies scholar Pat Gill has called "technostalgia"—the desire to revise the past to redetermine the present by harnessing technology toward human ends, all the while recognizing the impossibility of the endeavor (a common theme in science fiction films, like *The Terminator* series).[3] The claim might be said to embody a collective psychological need to believe that the nation was not as vulnerable as it appeared, that U.S. technological sophistication remained intact and in fact would have stopped the men had it been in place. This technostalgic longing to revise the past—a nostalgic sensibility wrapped up with an almost gleeful futurism—provides a paradoxical sort of origin myth for facial recognition technology. In the post-9/11 context, the technology emerged as an already existing, reliable, and high-tech solution to the newest, most pressing problem facing the nation. This move effectively elided the history of this technology, even as it inserted it fully formed into the past. In reality, well before 9/11 a whole set of social actors was already engaged in ongoing struggles and negotiations over the development and use of automated facial recognition technology. While it was *not* already fully formed and ready to identify the nation's post–Cold War enemy Other, it was already "embedded in and shaped by a rich web of cultural practices and ideas."[4]

This book explores the web of cultural practices and ideas, along with the policy programs and institutional priorities, in which automated face perception technologies are embedded. I investigate the effort, underway since the 1960s and gathering momentum since the 1990s, to teach computers to "see" the human face—to develop automated systems for identifying human faces and distinguishing them from one another, and for recognizing human facial expressions. This effort is best understood not as a unified program but as an interdisciplinary field of research and set of technological experiments. It is part of the broader effort to automate vision—to create machines that not only can generate images, but also analyze the content of those images. Computer scientists are interested in developing automated face perception technologies primarily as a means of creating more intelligent machines and more sophisticated forms of human-computer interaction.[5] Other social actors—especially military, state security, and law enforcement agencies—have viewed these technologies as uniquely suited to the development of "smart" surveillance, monitoring systems that perform the labor of surveillance with less human input and less need to rely on the perceptual capacities and analytical skills of human beings, with the possibility of creating new divisions of perceptual labor between humans and computers. New forms of human-machine integration promise to make surveillance systems function more effectively and extend their reach over time and space. But whether these experimental technologies can or should be made to accomplish these goals remains open to debate, one that often plays out in press and policy discussions as a trade-off between "security" and "privacy."

The purpose of this book is to move beyond the security versus privacy debate and make more adequate sense of the combination of technological projects that aim to create machines for identifying human faces and facial expressions, to understand the constellation of social forces that are shaping these technologies, and to consider what interest in their development and uses tells us about the society in which they are embedded. Why the perceived need to automate the human capacity for facial recognition and expression interpretation at this historical juncture? How are particular social problems conceptualized such that these technologies are posited as potential solutions? What forms are these technologies taking, and what is at stake in their development and institutionalization? In framing the main questions of the book in this way, I take my cue from communication scholars like Raymond Williams, James Carey, Susan Douglas, Carolyn Marvin, and Jonathan Sterne, who have made the unassailable case that technologies are thoroughly cultural forms from the outset, embodying the hopes, dreams, desires, and especially the power relations and ideological conflicts of the societies that produce them. My main focus is the United States, but it would be a mistake to view automated face perception technologies as uniquely "American." Research and development has occurred in other countries as well, and the technologies are finding applications beyond U.S. borders. But the spaces along and within the U.S. borders have become special sites of experimentation for automating the deceptively complex processes of facial identification and expression recognition. U.S. borders, airports, cities, suburbs, shopping malls, bars, casinos, banks, schools, workplaces, homes—and the bodies and behaviors of the inhabitants of these spaces—are special areas of experimentation for new surveillance technologies, in no small part because contemporary U.S. culture exhibits an intense preoccupation with the combined priorities of security and technology. Automated face perception technologies promise to provide high-tech security for the spaces and bodies of value in late capitalist societies, and belief in their ability to fulfill that promise plays a fundamental role in their institutionalization.[6]

Scholars writing in the field of surveillance studies have offered diagnoses about the preoccupation with security in late modern societies. The social theorist David Lyon is perhaps the strongest advocate of the view that these are essentially "surveillance societies," characterized by pervasive forms of social sorting and orchestration, and that surveillance should be taken seriously as an ethical, moral, and political concern.[7] A central political concern in debates about surveillance is whether the obsession with security and spread of new monitoring technologies are ushering in oppressive totalitarian societies akin to George Orwell's dystopic vision in *1984*. In his foundational study of five mass surveillance systems in Britain and the United States in the late 1960s and early 1970s (both state and commercial systems), the sociologist James Rule examined the extent to which each system approximated the "total surveillance society" depicted in Orwell's novel. He found that they shared "many of the sociological qualities . . . though of course they are much less powerful and they do not necessarily pursue the same malevolent purposes."[8]

Rule conducted his seminal study on the cusp of computerization, providing an invaluable snapshot of the moment and identifying many of the shortcomings in mass surveillance systems—in terms of size, degree of centralization, speed of information flows, points of contact with clientele, and capacity or effectiveness—that computer databases and networks promised to overcome. Subsequent studies have examined the influence of computerization on monitoring practices, including David Burnham's work on the rise of the "computer state," Oscar Gandy's examination of the computer-assisted market research machine (the "panoptic sort"), and Mark Poster's analysis of the intensified panoptic functions of the database form.[9] Much of this scholarship suggests that the novelty of these new computerized forms of surveillance derives from the extent to which computers have enabled distinct surveillance systems to converge into a larger "information infrastructure" or "surveillant assemblage."[10] Networked systems allow data to flow from one site to another, back and forth between state and private-sector organizations, enabling ever more sophisticated and effective forms of social control. These scholars maintain that, while late capitalist societies may not precisely mirror Orwell's vision, computerization is nevertheless enabling significant advancements in institutionalized forms of surveillance.

These arguments clearly have a basis in real developments. But a problem with theoretical claims about the convergence of surveillance systems afforded by computerization is a tendency to gloss over the amount of effort that goes into developing and integrating new technologies and systems.[11] The introduction of new surveillance technologies and the convergence of surveillance systems do not happen seamlessly or automatically, instead presenting major logistical, technical, and political challenges involving conflicts and negotiations among various vested social actors. In the case of automated facial recognition systems, a common perception is that these technologies are either already deployed in a variety of settings, or that their deployment is happening at a rapid pace. One encounters these assumptions repeatedly not only in industry discourse and press accounts but also in the surveillance studies literature, and even those who recognize that the technologies are not widely diffused tend to see them as a largely inevitable part of the not-too-distant future.[12] But while experimental systems have already been deployed in a variety of contexts, the widespread use of these technologies has never been a foregone conclusion. For reasons that I explore in this book, whether in fact automated facial recognition and expression analysis systems can accomplish what their proponents aim to accomplish remains an open question. Although developers are making incremental improvements in algorithms and other dimensions of software and hardware development, so far these technologies do not work very well outside constrained settings. Computerized face perception is proving to be an incredibly difficult technology to engineer.

The problem with assumptions about the rapid diffusion and integration of new surveillance technologies is not only that the process is more complicated than it seems. Understanding the experimental status of the technology is critical, because the prevalent myth of inevitability surrounding this and other new forms of surveillance itself performs an important role in their institutionalization, and in the broader effort to shape the future toward certain ends. Making authoritative predictions about increasingly ubiquitous and intrusive surveillance techniques encourages public acquiescence, while suppressing alternative, less technocratic ways to address complex social problems and envision a better future. Assumptions about the rapid development and convergence of surveillance systems support what William Bogard calls "the imaginary of surveillant control"—a hyper-real vision of perfectly functioning and totalizing surveillance that is more real than the real in the Baudrillardian sense.[13] Creating the illusion creates the reality, as technologies of simulation supersede the material forms and effects of actual monitoring systems. In short, people come to believe in the power and sophistication of surveillance systems, and this belief itself has important effects on social organization and practice. For this reason, according to Bogard, it is important to understand not so much "what surveillance is and does"—what a surveillance system is actually capable of—but instead how the image of totalizing surveillance itself informs the logic of system development and functions as a form of social control.[14]

While Bogard is right that the surveillant imaginary is itself a problem in need of critical analysis, this does not mean that social theories of surveillance can abandon consideration of the actual capacities of surveillance systems (leaving it to legal and policy analysts, as Bogard suggests). For social theory no less than legal and policy analysis, it remains crucial to understand "what surveillance is and does," to leave the reality principle intact rather than assuming that what really matters is the surveillant imaginary. Instead of making assumptions about the convergence of surveillance forms or the effects of a totalizing image of surveillance, it is important to investigate precisely how and to what extent system convergence is happening, and how beliefs about the power of surveillance help to forward the aims of developing more sophisticated and ubiquitous surveillance systems. While there is clearly a need to make theoretical speculations about the convergence of surveillance systems, there is also a need to understand precisely how surveillance technologies are developed, how convergence happens, and whose interests are served in the process. Why are certain types of technologies developed and certain systems integrated in the first place (to serve what specific needs or priorities)? What policies and procedures are put in place to make system integration possible? When and why does convergence fail, and what are productive effects of those failures? When are new technologies and efforts at system convergence effective and how is that effectiveness determined? "Technologies are not 'effective' by virtue of some straightforward appeal to science," write Haggerty and Ericson. "Instead, a determination of effectiveness is the net outcome of often contentious political struggles."[15] Although proponents of facial recognition and expression analysis systems often define their effectiveness in narrowly technical terms, in reality, determinations of their effectiveness involve contentious political struggles and intense persuasive efforts.[16]

By focusing on the politics of developing and deploying specific new technologies, this book aims to demonstrate not the inevitability of a particular technological future, but its profound contingency and contestability. In order to understand the social implications and historical significance of new computerized forms of surveillance, it is especially important to avoid making determinist assumptions about the inevitability of new technologies, and to understand the process by which proponents attempt to establish their authority, legitimacy, and necessity. I investigate what proponents want to achieve with automated face perception technologies and what would be required to put that vision into place. In the process of examining what would be necessary to bridge the gap between the current state of the technology and the vision, I identify the tensions and contradictions that would have to be designed into these systems—in other words, the ambiguity and error that the technologies would have to absorb in order to function effectively in real-world contexts. These inherent tensions and contradictions call into question claims about the accuracy and authority of new automated forms of face perception, and also raise questions about their effectiveness as technologies of security.

Drawing insights from communication scholarship, this book examines how the interrelated priorities of institutional actors are shaping the effort to program computers to identify human faces and facial expressions, including business entities, law enforcement, and state security agencies. Work on the history of media and communication technologies has much to offer the study of new surveillance technologies, in no small part because these technologies are themselves forms of media that both derive from and help to shape the evolution of technologies like video, television, telephony, and computing. The drive to produce more effective forms of surveillance has had a consistent role to play in the development of communication technologies, including visual media technologies.[17] As Tom Gunning has shown, a specific and direct historical relationship exists between the physiognomic analysis of the face and the development of photography and motion pictures. "The desire to know the face in its most transitory and bizarre manifestations was stimulated by the use of photography," writes Gunning, "but that desire, in turn, also stimulated the development of photography itself, spurring it to increasing technical mastery over time and motion, prodding it toward the actual invention of motion pictures."[18] A cultural analysis of automated facial recognition and expression analysis technologies provides evidence that the drive to "know the face" continues to be stimulated by new photographic technologies, while at the same time pushing the development of these technologies in particular directions.

As technical projects, automated facial recognition and automated facial expression analysis pose distinct problems, and they represent related but somewhat distinct research and development efforts. Strictly speaking, facial recognition technology treats the face as an index of identity, disregarding its expressive capacity and communicative role in social interaction. This is nowhere more apparent than in new rules prohibiting drivers from smiling for their driver's license photos, in order to improve the accuracy of computer matching.[19] The aim is to use the iconicity of facial images as a means of establishing their indexicality, their definitive connection to real, embodied persons. Automated facial expression analysis, on the other hand, targets precisely what facial recognition technology attempts to control for—the diverse meanings that an individual face can convey—and in this way promises to accomplish what facial recognition technology fails to do, using the surface of the face to see inside the person. Where automated identification of individual faces disregards their affective qualities, automated facial expression analysis treats those affective dimensions as objects for precise measurement and computation.

However distinct these projects are in a technical sense, the face is at once both a marker of identity and a site of affect, and the differentiation of the face along these lines serves the purposes of scientific investigation and technical engineering more than it does a theoretically rich understanding of human communication. The diverse range of ways human beings use their faces and interpret the faces of others does not reduce easily to a set of technical processes. In a basic sense, the face is never a static object, and different facial expressions can change the appearance of the face considerably. Not surprisingly, there is also overlap in technical efforts to develop automated forms of facial identification and expression analysis. The initial steps of automated facial recognition—"face detection" and "feature extraction," or locating a face in an image and extracting relevant features—are also the necessary first steps in automated facial expression analysis. In addition, computer scientists have investigated individual differences in facial expressiveness as a means of augmenting the accuracy of facial identification algorithms.[20] More significantly, the possibility of robust computer vision systems depends on the fusion of technical systems designed to simulate these interrelated forms of visual perception. While there may be distinct applications for identification versus expression analysis, at least in the short term, the predominant view in computer science is that more advanced forms of computer vision require designing systems that can adequately perform both of these functions, and especially to do so as well as, or preferably *better than,* humans.

The question of whether computers can be made to identify faces and facial expressions as well as or better than humans raises philosophical questions concerning the nature of sight and visual perception that have animated theoretical debates about both artificial intelligence and visual media technologies. The question at the heart of the Turing test, about whether a machine can "think" or exhibit humanlike intelligence, focused on the ability of a computer program to convincingly manipulate natural language, but a similar question likewise pertains to whether a machine can "see." The answer of course depends on what it means to see, and while "seeing" is obviously a physiological process, it is also a cultural practice, shaped by social and historical forces. The art historian Martin Jay uses the term "scopic regime" to describe an ideal typical combination of visual theories and practices that together create more or less unified "ways of seeing."[21] While there may be multiple, competing scopic regimes operating in any particular context, most theorists of visual culture agree that the dominant scopic regime of modernity is what Jay calls "Cartesian perspectivalism," the privileging of a rational observing subject capable of surveying the world objectively and from a distance, in its totality. Kevin Robins has extended this concept to an analysis of new visual imaging technologies, arguing that they embody the characteristic Cartesian desire for visual sovereignty, signaling the progressive rationalization of vision in order to establish mastery and control over a chaotic world.[22] As these arguments suggest, sight and vision should not be understood as essential qualities or strictly physiological processes with universal, transhistorical meanings or functions. There is no such thing as natural or "true" vision. Instead, "vision is always a question of the *power to see*," and "struggles over what will count as rational accounts of the world are struggles over *how* to see," as Donna Haraway has argued.[23] Because seeing is as much a cultural practice as a physiological process, Suzannah Biernoff notes, "it cannot provide a historical common ground or the basis of a shared aesthetic experience."[24]

It follows that *if* machines can see, they must necessarily embody particular ways of seeing, rather than possessing a universal, disembodied, objective form of vision, outside of any particular vantage point or subject position. This is as true of automated face perception as it has been of other photographic technologies. The *digitizing impulse* behind the development of computer vision techniques is in one sense a re-articulation of the *mechanizing impulse* that motivated the adoption of photography by the positivist sciences in the nineteenth century, as Daston and Galison have documented—another way of standardizing images and eliminating individual human judgment in their interpretation.[25] Surveying the world from afar in a detached, authoritative way is the goal that underpins most machine vision projects.[26] In discussions about the merits of automated face perception technologies, one repeatedly encounters claims to their authority and technical neutrality. Simply by nature of being computerized, facial recognition systems are deemed more accurate and objective and less subject to the prejudices and apparent inadequacies of human perception. Automated facial expression analysis similarly promises to introduce a level of precision to the interpretation of facial expressions, detached from and exceeding human perceptual capabilities. The claim is that these technologies will create accurate, precision-guided, objective, all-seeing machines that function much better, more efficiently, and more powerfully than human perception alone.

This book unpacks such claims by examining what is involved in the effort to invest computers with the capacity to recognize human faces and facial expressions. Quite a bit of effort goes into this project: the painstaking development of algorithms to digitize the analog world of faces, the amassing of vast storehouses of facial images to serve as the basis of computerized visual memory, the development of efficient image retrieval methods, the tagging of billions of images with metadata to make them more searchable, the wiring of physical spaces with networks of cameras and computers and other hardware, the development of software interfaces that make sense to human users of these systems, the training of people to use the technologies and continuous retraining to help them keep pace with software and hardware upgrades, and much more. Peering inside the construction of this massive machinery of vision reveals that computers "see" only in a metaphorical sense, only in highly constrained ways, and only with a significant investment of human effort. Computer vision systems are very much constrained by the purposes of their design, and suggesting that a computational model of vision represents an objective, detached form of vision elides the intentions and labor behind the design, deployment, and uses of these technologies. The desire to overcome the ambiguities and interpretive flexibility inherent in human perception and social interaction is driving the development of incredibly complex machines that can do what humans do everyday but in a totally imperfect way. These new techniques of observation will not lead to the fulfillment of the rationalist program, as Kevin Robins suggests. The rationalist program will never be complete, but the longing to fulfill it, and the vain belief that it can be fulfilled—that total, perfect knowledge of the world is possible—is one of the major motivating forces behind the pursuit of new technologies of vision and new human-machine visual assemblages.

Just as there is no standard or universal way of seeing, there is no universal way of seeing the face. A significant contingent of psychologists and other researchers who study the face suggest otherwise, but I take the position in this book that there are many ways of seeing the face, and that the meaning and experience of face-to-face interaction is a historically and culturally variable set of practices. We find evidence of this variability, for example, in research suggesting the existence of an "other race effect" or an "own-race bias" in people's ability to recognize faces. People tend to have more difficulty recognizing the faces of people outside their primary social group, meaning that individuals learn to see some faces better than others. Humans' capacity for facial recognition is a specialized area of research in fields like psychology and neuroscience, and much of this research suggests that there may indeed be something physiologically special about the way humans see and recognize faces that distinguishes this practice from the way they see and identify other types of objects.[27] But even the obvious existence of a physiological dimension to face perception does not mean that there exists a universally human way of seeing the face. While face perception may in fact be "the most developed visual perceptual skill in humans,"[28] the critical role that it plays in human social interaction—and the wide array of technologies and practices designed and implemented to facilitate facial representation and interpretation—means that face perception cannot help but vary widely in different contexts. The ways humans see the faces of others change, of necessity, along with changing cultural practices, social conventions, and forms of social and technological organization. In short, face perception does not reduce to a universal physiological process that can be manifested in a standardized computational system; instead, there are a wide variety of "scopic regimes" of the face, a wide range of ways in which people use their faces and interpret the faces of others. Computation is itself a culturally and historically specific way of analyzing faces and modeling visual perception.

Facial recognition technology is more advanced in its development and applications than automated facial expression analysis, and for this reason receives considerably more attention in this book. Computer scientists began developing algorithms for detecting faces in images and distinguishing faces from one another in the 1960s, as part of the range of problems being addressed in the areas of automated reasoning and pattern recognition. Not surprisingly, the Department of Defense funded much of the research in the United States, marrying the automation of facial recognition to military priorities since its inception. After several decades of sporadic development, limited to some extent by the available computing technology but also by the lack of well-formulated social uses, prototype systems began to take shape. New companies with names like Visionics, Viisage, and Miros Inc. started marketing commercial facial recognition systems in the 1990s. The early period of commercialization—still ongoing—has involved the search for markets for unproven products, with proponents working to move the technology beyond military applications. Vendors have directed most of the effort toward building a customer base among institutional users—military as well as civilian government agencies, police departments, and business enterprises, including manufacturers of computers, automated teller machines, and other electronic equipment—working with these potential customers to define compelling social needs for more high-tech forms of surveillance and identification in the form of expensive, large-scale, database-driven facial recognition systems.

In its applications for biometric identification, facial recognition technology is one of array of technologies being developed to address a fundamental concern of modern societies: the problem of "disembodied identities," or the existence of visual and textual representations of individuals that circulate independent of their physical bodies. Long before the development of audio-visual media and electronic databases, the circulation of visual and textual representations created the conditions whereby certain classes of human identities became unmoored from their bodily existence. But it was the new communication technologies of the nineteenth century, like telegraphy, photography, telephony, and the phonograph, that gave these representations new mediated forms and amplified the uncanny phenomenon of ubiquitous incorporeal replicas moving through society disarticulated from their embodied human counterparts. In 1886, Frederic Myers, a member of the British Society for Psychical Research, coined the phrase "phantasms of the living," to refer to the proliferation of these humanoid replicas.[29] "What men and women in the late nineteenth century faced with alarm," writes John Durham Peters, "is something we have had over a century to get used to: a superabundance of phantasms of the living appearing in various media."[30] While we are now quite accustomed to the existence of these "phantasms of the living" and the representational roles they play in our lives, the period of computerization has seen a renewed explosion in their quantity, forms, and uses, intensifying the problem of how to connect them back to embodied persons. One of the main reasons why digital biometric technologies are taking institutionalized forms at this historical juncture is because they promise to resolve one of the central problems of communication in large-scale societies: bodies missing in action from mediated communicative contexts.[31]

Insofar as they promise to re-embody disembodied identities, new biometric forms of identification supplement, and even replace in some cases, what Craig Robertson calls a "documentary regime of verification."[32] This system of standardized documents, archives, and administrative procedures for the management of individual identities itself displaced the more personal and informal forms of trust and recognition characteristic of smaller-scale forms of social organization. The aim of a documentary regime of verification was to assign each individual an official identity that could be verified in repeated transactions with the state and other institutions. These official forms of bureaucratic identification cobbled together a set of existing and already mediated markers of identity—such as names, addresses, signatures, and photographs—to create a more stable and standardized form of identity that could be verified via the very bureaucratic apparatus that constitutes that identity. In short, our seemingly self-evident "official identities" are in reality a product of bureaucratization and a relatively recent historical construction, and considerable effort has gone into designing systems that can produce and reproduce these identities.[33]

Much like the drive to produce more advanced and precise forms of visual perception, states and other institutional users of identification systems are in constant pursuit of the perfect means of identification. Each new innovation in identification systems (fingerprinting, standardized documents, ID photos, computerized record keeping, and machine-readable documents, to name a few) has been designed to resolve some of the tensions inherent in the mediated process of identification, and especially to lay claim to making an absolute, immediate connection between bodies and their official identities. The aim of biometric identification technologies—like optical fingerprinting, iris scanning, and voice recognition—is to bind identity to the body using digital representations of unique body parts, or, in the case of voice printing, by capturing, digitizing, and analyzing the sounds that the body produces. Digitization—translating images and other analog texts into binary code—is the latest in a long line of techniques posited as a definitive, accurate, and objective means of binding identity to the body. The claim is that by digitizing visual representations of the body, the body itself will be laid bare and tied directly into information networks.

Today, DNA in particular is viewed as the ultimate identifier, the most precise and scientific means of linking bodies to identities. Our genes are thought to be quite literally the code that determines who we are, to represent our identities in an absolute sense. But genes themselves are not codes that identify bodies; they must be translated into coded form using a specialized technique of inscription, a process of mediation that involves layers of technical integration. A genetic code ostensibly makes a direct link to physical bodies when in fact it does no such thing. Not only does it require a process of technical mediation to generate, but a genetic profile establishes only a probability. It cannot point definitively to a specific body, and it cannot serve as an index of identity unless it is connected to other identifying information about the person it represents. Belief that a genetic code is literally extracted from the physical body (and thereby connected to it in an unadulterated, absolute sense) is a form of what Donna Haraway calls "gene fetishism." Drawing on Alfred Whitehead's notion of "the fallacy of misplaced concreteness," Haraway explains that gene fetishism mistakes the abstraction of the gene for a concrete entity: "A gene is not a thing, much less a 'master molecule,' or a self-contained code. Instead, the term *gene* signifies a node of durable action where many actors, human and nonhuman, meet."[34]

The claim to authority of biometric identification rests in part of on a similar form of fetishism, a disavowal of the mediated relationship between bodies and identities and the overdetermined process by which a "biometric" is produced. In laying claim to a direct link to bodies, biometric technologies promise to stabilize the messy ambiguity of identity, to automatically read a stable, individual identity directly off the body. To be effective, the connection that biometric technologies establish between identity and the body must appear natural and self-evident: *of course* our true identities can be ascertained by scanning and digitizing our faces, eyes, voices, hands, and fingertips, the logical next steps in identification systems. But the assumption that biometrics are derived from and link directly to physical bodies conceals a complex technological process of mediation, as well as a whole set of historical relationships that create the very conditions of possibility for biometric identification.

In her work on protocols of identification in nineteenth-century Europe, Jane Caplan explains that any standardized identification system depends for its effectiveness on the stability or replicability of its operations, and their relative invisibility (i.e., their naturalization).[35] Stable, standardized identification systems were a crucial development for societies increasing in size, mobility, and anonymity. But "in spite of every effort at stabilization," writes Caplan, "the culture of identification was—is—essentially unruly, not only because of its vast scope and variety, but because even in its most controlling and technologized forms it is based on a concept that is itself difficult to stabilize and control"—the concept of *identity*.[36] The term "identity" signifies both what is unique about an individual and what associates her with others like her, a dual meaning that keeps the concept "constantly in motion against itself, even before one addresses any of the mechanical problems of operating an efficient system of identification in practice."[37] The inherent instability and mediated characteristics of identity make building identification systems a process fraught with difficulty, and these challenges have in turn led to perpetual efforts at standardization. The turn to digital biometric identification represents the latest in a long line of efforts to stabilize and standardize identification systems, and to push those standardized categories of identity back out onto individual bodies.

In its application to identification systems, the dual aim of facial recognition technology is to automate the mediated process of connecting faces to identities, and to enable the distribution of those identities across computer networks in order to institutionalize a more effective regime of mass individuation than afforded by documentary identification alone. By *mass individuation* I mean the process of extending across an entire population technologies and procedures for treating each individual as a specific case, a process greatly facilitated by computerization and the development of networked databases.[38] Mass individuation supports other social processes and political-economic priorities, such as *mass customization*, or the mass production of customized goods and services, and the *individualization of labor*, the process of identifying and individuating workers in order to measure their precise contribution to production.[39] Mass individuation is also a modern governmental strategy for security provision and population management, a social regulatory model that involves knowing in precise detail the identity of each member of the population in order to differentiate individuals according to variable levels of access, privilege, and risk. In other words, the possibility of digital biometric identification should not be understood in narrow terms as the natural outgrowth of technical advancements in identification systems. Instead, these technologies are being envisioned and designed to fulfill certain perceived social necessities and political-economic demands of large-scale, late capitalist societies—societies characterized by a predominance of mediated forms of social organization and vastly asymmetrical distributions of wealth. The expansion of computer networks has created new problems of communication-without-bodies, necessitating new techniques for identifying people, verifying their legitimate identities, and otherwise gaining knowledge about who they are.

But while facial recognition and other biometric technologies hold out the promise, they can never completely resolve the inherent problems with building stable identification systems for the mass individuation of large-scale societies. Digitizing faces, storing infinitely reproducible images of faces in databases, networking those databases, and designing more sophisticated image retrieval techniques are technical practices that can increase the scale of identification systems. Automation facilitates the standardization of identification systems and the expansion of their reach over time and space. Facial recognition technology in particular promises to provide a means for identification "at a distance," in terms of both the measurable distance between cameras and people in particular local spaces, and the more extended reach of automated identification over networks covering distant locations. But the automation of facial identification cannot definitively stabilize identity. No matter how sophisticated the algorithms for matching facial images or how comprehensive the image databases, the tension remains between identity "as the *self-same*, in an individualizing, subjective sense, and 'identity' as *sameness with another*, in a classifying, objective sense."[40] What Stuart Hall has argued about "cultural identity" is in fact also true of the official, bureaucratic form of identity: it is not a fixed and stable object or "an already accomplished fact," but a "'production,' which is never complete, always in process, and always constituted within, not outside, representation."[41]

This does not mean that the automation of facial recognition is a meaningless endeavor, bound to fail as a means of establishing consistently identifiable bodies over time and space. There will be technological failures, to be sure, but the technology is in many ways already effective, because the very pursuit of computerized, biometric forms of identification suggests that we are witnessing a reconfiguration of identity, the attachment of new meaning and new practices to what identity is and to how it works. Identity remains that which both differentiates us and associates us with others. But we now have new ways of identifying ourselves and being identified that did not exist before, and there is an increasing quantity of instances where we are required to establish our identities definitively so that our status or level of access can be determined, and information about those transactions can be recorded, with all that data in turn becoming part of our identities.[42] Identity is now understood as a disembodied aggregate of data, a digital representation of the person constructed over time and space based on the perpetual collection of more data. Although difficult to measure empirically, it is hard to deny that people living in modern societies today engage in more transactions on a daily basis that require them to interface with identification systems than people did in the past. Rather than stabilizing identity, the proliferation of these transactions is making it messier than ever, in turn leading to the perpetual pursuit of new efforts at stabilization. It is difficult to know where this seemingly self-fulfilling process is going, but it would certainly not be insightful to say that the practices of identity remain the same as they ever were.

Facial recognition technology should not be conflated with other types of biometrics. It is unique relative to other forms of biometric identification because the content of the medium is the image of the human face. This makes it an especially challenging technical problem and puts it at a disadvantage relative to other biometrics in terms of its level of development, ease of adoption and use, and general viability. Optical fingerprinting is in many ways a more accurate and reliable means of binding identities to bodies, for example. As two MIT researchers put it, "Developing a computational model of face recognition is quite difficult, because faces are complex, multidimensional, and meaningful visual stimuli."[43] An individual face changes considerably not only with its surface movements, but also with aging, trauma, surgery, makeup, and lighting. Faces themselves change over time, and images captured of a face can be of highly variable quality. The variability of faces across populations, as well the dynamic states of the individual and the range of images that can be rendered of a particular person, make automated facial recognition a very challenging technical problem.

In short, like identity, the face is a difficult object to stabilize. Computer scientists have developed a variety of different techniques designed to translate an image of the face into a "facial template," a smaller amount of data that can be compared against existing images stored in a comparison database.[44]

And the digitization of facial images is only one small part of the design of facial recognition systems. Faces must be detected in images, extracted from the background, and "normalized" so that they conform to a standard format. The matching process typically results in not one but a range of possible matches, depending on a set "matching threshold." High matching thresholds increase the chances of missing a positive match, while low matching thresholds can produce a large number of "false positives." At the level of their applications, automated facial recognition systems are divided into two general types: those that use static images of the face and those that analyze dynamic images of faces from video.[45] Applications can also be differentiated according to whether the aim is to *verify* the identities of individuals (to determine whether people are whom they claimed to be, for example, at a border crossing station or when engaging in a financial transaction), or to *identify* people whose identities are unknown (in urban or crowd surveillance scenarios, for example). The first problem requires a one-to-one facial image comparison, while the second problem involves a more technically challenging and information-intensive process of comparing an image captured of an unknown person's face against a database of facial images (a form of content-based image retrieval, or CBIR).

While facial recognition technology presents considerable technical challenges, it also possesses certain advantages over other biometrics. One set of advantages involves the practical improvements it portends for identification systems; for example, "it poses fewer demands on subjects and may be conducted at a distance without their knowledge or consent."[46] Along with the practical upgrades it offers for surveillance and identification systems, the technology's use of the face as an object of identification invests it with certain cultural and ideological capital. Facial recognition technology combines an image of high-tech identification with a set of enduring cultural assumptions about the meaning of the face, its unique connection to individuality and identity (in its multiple, conflicting senses), and its distinctive place in human interaction and communication. It is for these reasons, as much as its practical advantages, that facial recognition technology has received special attention.

Long before computer scientists began developing techniques for automated face perception, visual media technologies were being developed and used to analyze, classify, and identify human faces. The use of the face as an object for the social and biological classification of people has a long and sordid history, indelibly tied to the use of the human sciences to justify social inequality. The science of physiognomy, widely accepted in the West until the mid to late nineteenth century, held that people's faces bore the signs of their essential qualities and could be visually analyzed as a means of measuring moral worth. Photography was invented at the height of this "physiognomic culture" and was put to use for facial analysis and classification. Most famously, the psychiatrist Hugh Welch Diamond (1809–1886) used photography to analyze and document the alleged facial indicators of insanity, and the eugenicist Francis Galton (1822–1911) developed a form of composite photography that he used to claim the existence of criminal facial types. As historians have shown, physiognomic classifications exhibited a consistent tendency to conflate ostensible facial signs of pathology with racial and class differences.[47] Galton's work in particular was "the culmination of all the nineteenth-century attempts to objectify, classify, and typify humans through portraiture," and Galton's ideas went on to form the basis of many of the racist, classist, and biologically determinist theories about human difference promulgated in the twentieth century.[48]

In what ways does the automation of face perception connect to this history? For its part, facial recognition technology appears to diverge radically from the business of using the face as an object for the social or biological classification of people. In practical terms, the aim of facial recognition systems is to identify individuals, to use the face like a fingerprint—as an index or recorded visual trace of a specific person. Computer scientists have taken a variety of approaches to developing algorithms for automated facial recognition, and the specific techniques devised for digitizing facial images are not necessarily based on assumptions about facial typologies. In technical terms, the development of algorithms for translating images of faces into digital "facial templates" is more or less divorced from the explicit social and biological classification of faces in the conventional sense. Developers of facial recognition algorithms have attempted to operationalize a more individualizing form of facial identification. They are not primarily interested in measuring differences in facial features between people of different racial or ethnic identities, for example. This does not mean that things like skin color and other facial signs of racial, ethnic, or gender difference are irrelevant to the development of these techniques, but the classification of faces according to race, ethnicity, or gender is not, by and large, the problem that computer scientists are trying to solve in their efforts to design automated forms of facial recognition.

The individualizing logic that underpins the development of facial recognition systems gives them more obvious genealogical ties to the system of anthropometry developed in the late nineteenth century by Francis Galton's contemporary, the Paris police official Alphonse Bertillon (1853–1914), than to Galton's photographic composites. As Allan Sekula has shown, Bertillon's system of anthropometry disregarded the possibility of generic categories and instead concerned the more practical, administrative aim of identifying individual criminals (much like fingerprinting, though significantly more laborious, since anthropometry involved not only taking photographs but also recording a series of bodily measurements and coded descriptions of each subject's body). Like anthropometric measurement, the algorithms developed to digitize faces and link those "facial templates" to identities are not designed to reveal anything about the essence of the individual. Unlike physiognomic analysis, facial recognition algorithms are not meant to derive knowledge of the interior of the person from the surface of the face. The individualizing logic that informs the design of facial recognition technology is part of the basis for claims about its technical neutrality—the aim is to identify individual faces rather than facial types.

But there are a number of problems with claims about the individualizing logic and technical neutrality of automated facial recognition. First, as I have already discussed, "identity" itself fundamentally embodies an individualizing and classifying logic, an inescapable tension that manifests at the level of system design. Even if the explicit classification of identities does not occur at the level of algorithm development, it does happen at the level of database construction and in the specific applications that give facial recognition technology a functioning form.[49] Again, facial recognition algorithms, or techniques for digitizing the face, represent only one part of the operation of facial recognition systems; most of these systems are designed to make use of an archive of facial images that define the parameters for the class of individuals that the system will identify. And forms of *social* classification, if not outright *biological* classification, inevitably happen at the level of database development, whether the database is a terrorist watchlist or an A-list of preferred customers. This was likewise true of the archival practices developed in the early application of photography to criminal identification. In Bertillon's system of anthropometric identification, Sekula explains, the camera alone was a limited technology. The contribution of photography to criminal identification came with its integration into a larger ensemble: "a bureaucratic-clerical-statistical system of 'intelligence.'"[50] A similar bureaucratic-statistical apparatus is required to make facial recognition algorithms effective for large-scale identification systems. "The act of recognition relies on comparison," writes Gunning, "and knowledge resides not in the single photograph, but within a vast photographic archives, cross-indexed by systems of classification."[51]

A related problem with claims about the technical neutrality of facial recognition technology and its divergence from the problematic assumptions of physiognomy has to do with its symbolic associations with antiquated notions of facial typologies. Facial recognition technology is inescapably tied to cultural assumptions about the relationship between the face and identity, including enduring beliefs about faces as carriers of signs that reveal the essential qualities of their bearers. As we will see in chapter 3, the technology's metaphoric connection to archaic ideas about physiognomic facial types gave it a special edge in the effort to define it as a security solution in the post-9/11 moment, especially through the trope of the "face of terror." The repeated use of this trope in press and policy discussions and in industry discourse about facial recognition technology did important ideological work, implying that embodied evil could be read off the faces of "terrorists," even if those faces had to be identified one by one, with the help of a new, methodically individualizing form of computerized facial identification. The racist implications of the "face of terror" trope underscored the underlying logic of social classification that informed the strategy of deploying new identification technologies for "homeland security." Although facial recognition algorithms were *not* designed to classify faces according to particular identity typologies, the deployment of large-scale, database-driven identification systems very much depended on, and promised to facilitate, a set of "biopolitical" security strategies that aimed to differentiate the population according to racially inflected criteria for determining who belonged and who did not, who was entitled to security and who posed a threat to that security.[52] Far from incorporating a neutral, all-seeing mode of visual perception, facial recognition systems promise to facilitate the diffusion of particular institutionalized ways of seeing, ones that rely on and seek to standardize essentialized identity categories under the guise of what appears to be a radically individualizing form of identification.

Both automated facial recognition and automated facial expression analysis gain metaphoric leverage in their historical connection to what Gunning refers to as "the semantically loaded and unceasingly ambiguous representation of the human face."[53] Both of these technological projects have genealogical ties to the history of using the face as an object of scientific investigation and social differentiation. But while facial recognition algorithms are not designed to use the surface of the face to reveal something about the interior of the person, automated facial expression analysis is another matter. The aim of automated facial expression analysis, or AFEA, is to peer inside the person, using the dimensions and intensities of facial movements as a means of determining what people are feeling and thinking. Where facial recognition technology treats the face as a "blank somatic surface" to be differentiated from other faces, AFEA treats the dynamic surface of the face as the site of differentiation—not a blank somatic surface but a field of classifiable information about the individual.[54]

The facial expression classification scheme devised by the psychologist Paul Ekman and his colleagues in particular promises to play centrally in the development of AFEA. Ekman is best known for his work on deception detection, the influence of which has extended beyond the academic field of psychology to the development of police and military interrogation techniques.[55] In the 1970s, Ekman and his colleague Wallace Friesen undertook an eight-year-long study of facial expressions, creating a scheme of forty-four discrete facial "action units"—individual muscle movements combinable to form many different facial displays. This "Facial Action Coding System," or FACS, is now the "gold standard" system for analyzing facial expressions in psychology, and as we will see in chapter 5, computer scientists see it as a promising approach for the automation of facial expression analysis. In turn, a host of social actors see computerized facial expression recognition as holding special potential to facilitate automated deception detection and other forms of affect analysis in a wide range of settings.

The stakes can be high in applying a system of classification and a computational logic to the human world of emotional or affective relations. Something inevitably happens to affect—to our understanding of what affect is and how it works—when it is treated as data to be processed by computers. One of the primary aims of FACS and FACS-based AFEA is to make human affective behaviors more *calculable*, to open them up to precise measurement and classification, thereby making them more amenable to forms of intervention, manipulation, and control. In this way, the technology promises to function like other forms of psychological assessment—as a technique of subjectification, a means of applying normalizing judgments to individual behaviors in order to shape or direct those behaviors in particular ways. Like earlier applications of photography to the analysis of the facial expressions, automated facial expression analysis is intimately connected to the social regulation of affect, and to the corresponding project of "making up people,"[56] marking out the parameters of normalcy and establishing a set of prescriptions for conduct, including the range of appropriate emotional responses to the world and the ways those responses are enacted through the face.

The face, like the body, does not exist somehow outside of history, and the very possibility of automating the human capacity for face perception gives us cause to consider the cultural reconstruction of the face in these times. What exactly do we mean when we refer to "the face"? The face is often conceived as the site of our identity and subjectivity, the source of our speech, the location of much of our sensory experience. But thinking about the face in these terms requires disarticulating it from the rest of the body as well as from its social, historical, and material context, slicing it off from the physical and discursive networks that allow it to take in air, food, and water; to express, blink, and sniffle; to acquire and produce signals; and to otherwise engage in meaningful exchange with the world. Faces are assemblages of skin, muscle, bone, cognition, emotion, and more. But certainly we have faces, however inseparable they are from our bodies and the worlds we inhabit. We know what faces are and what they do, and we learn to use and interpret them in particular ways. We are now perfectly comfortable metaphorically cutting them off from our bodies by photographing them and treating those images as objects-in-themselves. Our photographed faces do not diminish our subjectivities, our identities, or our relations with others. Rather, photography is now used as a means of constructing, facilitating, and enhancing these dimensions of what it means to human. Of course it is also used to classify and individuate us in ways over which we have no control, and these different impulses of photography—its honorific and repressive tendencies, to borrow Allan Sekula's terminology[57]—continue to intersect in complex ways along with the development of new photographic techniques and practices.

Photographic and archival practices have helped create a false unity of the face—the very idea of the face as a singular, unified object, detached from the body and from the world. The idea that we might think of "the face" as an object-in-itself begins to fall apart when we see faces as assemblages, and when we consider the vast range of differences among faces, what they look like, what they do, and especially what they mean, within cultures, across cultures, and over time. Deleuze and Guattari claim that "all faces envelope an unknown, unexplored landscape," by which they surely mean something profound.[58] In "Year Zero: Faciality," they argue that concrete faces do not come ready made but are produced by an "abstract machine of faciality," a process of "facialization" whereby bodies and their surroundings are reduced to the face, a "white wall, black hole system."[59] The face that is produced by the abstract machine of faciality is an *inhuman* face, and facialization can be a violent process, transforming the radical and deeply relational potential of becoming into a dichotomous relationship between signifier and signified.[60] The paradigmatic instance of facialization is the close-up of the face in cinema, which destroys what is recognizable, social, and relational about the face and "turns the face into a phantom."[61]

Although precisely what Deleuze and Guattari mean by the abstract machine of faciality is open to interpretation, one imagines that the translation of an embodied face into a digital "facial template," the circulation of millions of facial templates over networks, or the meticulous classification of facial movements according to a standardized coding system would fuel the engine of such a machine. If an abstract machine of faciality actually exists, then the automation of facial recognition and expression analysis would have to count as part of that machine's "technological trajectory." In his classic study of nuclear missile guidance systems, Donald MacKenzie uses the term "technological trajectory" to refer to an institutionalized form of technological change, a course of technical development that appears natural and autonomous from a distance because it has a relatively stable organizational framework, because resources are channeled to support the activities of that framework, and because the prediction that the technology can be made to work is perceived as credible.[62] In other words, a "technological trajectory" is a sort of "self-fulfilling prophecy."[63]

This book documents efforts to build a stable framework for the development of automated face perception technologies and to channel resources to support that framework. It also examines efforts on the part of proponents to make a credible case for the viability and desirability of these technologies. Functioning face perception technologies depend, now and in the future, on the formulation of compelling social uses and on a measure of faith in the capacity of the technology to do what it purports to do, regardless of whether it can do so with complete accuracy. The ability of computer scientists and other social actors to design machines that can "see" human faces depends on whether people believe that it is both possible and useful. Automated facial recognition and facial expression analysis do not have to work perfectly to be effective, and our belief in the inevitability of these technologies has important effects on their development and their uses. To take shape as functioning technologies, automated facial recognition and expression analysis must be viewed as obvious, necessary, and inevitable next steps in the technological trajectory of facial representation, identification, and interpretation—in the abstract machine of faciality.

# 1

# Facial Recognition Technology from the Lab to the Marketplace

At the 1970s World's Fair in Osaka, Japan, the Nippon Electric Company (NEC) staged an attraction called "Computer Physiognomy." Visitors to the exhibit would sit in front of a television camera to have their pictures taken and then fed into a computer where a simple program would extract lines from the images and locate several feature points on their faces. In one last step, the program would classify the faces into one of seven categories, each corresponding to a famous person. As the computer scientist Takeo Kanade wrote in his doctoral dissertation, "the program was not very reliable," but "the attraction itself was very successful," drawing hundreds, probably thousands, of people to have their faces scanned and categorized in this new, state-of-the-art manner.[1]

NEC's largely forgotten World's Fair attraction was part of the highly successful Osaka Expo '70 that showcased Japan's postwar technological accomplishments and rapid economic growth.[2] The exhibit, like other attractions at World's Fairs before and after, clearly aimed to marry technology with amusement, offering an "almost visceral pleasure. . . to sell the comparatively cold, abstract, and at times unappealing accomplishments of the technological."[3] It also married the archaic with the futuristic. It would have been naïve to believe that the program actually produced scientific results, since physiognomic analysis had long since been discredited as pseudoscience. But the experience of having one's face analyzed by a computer program foretold a future in which intelligent machines would see the human face and make sense of it in a technologically advanced way. As Langdon Winner has observed, World's Fairs have historically offered a vision of the technological future as an already accomplished fact, giving visitors a sense of being propelled forward by forces larger than themselves. "There were no pavilions to solicit the public's suggestions about emerging devices, systems, or role definitions," instead promoting a model of closed, corporate-sponsored research and development of which ordinary people need only be in awe.[4] It is no stretch to suggest that the NEC Computer Physiognomy attraction conveyed this implicit message. One can imagine that in 1970, the experience of having a computer scan and analyze one's own face might have impressed on the sitter a powerful sense of her own future subjection to information processing machines. At the very least, it demonstrated one of the many uses for which computers would be put: sorting human beings into distinct social classifications. It was hardly a new idea, but the computer seemed to give it a whole new edge.

The Computer Physiognomy exhibit also had a more direct use value, however, yielding a database of facial images for budding research on the problem of computer face recognition, images of "faces young and old, males and females, with glasses and hats, and faces with a turn, tilt or inclination to a slight degree."[5] Kanade used 688 of the photographs for his dissertation research in electrical engineering at Kyoto University in the early seventies, which built on early work in pattern recognition and picture processing to devise a facial feature extraction program. (Kanade would eventually become the head of the Robotics Institute at Carnegie Mellon University.) Many of the civilian and academic computer scientists working on the problem of computer image processing in these early days were motivated by the prospect of creating more intelligent machines, part of dedicated research teams working on the "quest" to create artificial intelligence.[6] The earliest research on computer recognition of human faces was one small part of the research programs in computer vision and robotics, themselves arms of artificial intelligence research.[7] In the United States, computer scientists working in the area of artificial intelligence made advances in computing technology—in computer vision as well as speech recognition and other areas—thanks in no small part to an enormous amount of military funding in the postwar period.[8]

But computer scientists and military strategists were not the only social actors with an interest in the technology. Computerized face recognition and other forms of automated bodily identification soon came to the attention of other social actors in both the public and private sectors who were recognizing an intensified need to "compensate for lost presences"—to deal with the problem of a proliferation of disembodied identities residing in databases and circulating over networks.[9] The late 1960s was the beginning of the expansion of computer networking in the business and government sectors, and businesses like NEC saw even more expansive growth ahead.[10] Indeed, electronics manufacturers, telecommunications companies, and other businesses were about to invest considerable effort into making it happen. As Dan Schiller has documented in painstaking detail, the big business users of expanding global computer networks spent the last three decades of the twentieth century laying proprietary claim to the network infrastructure, jettisoning many of the public service tenets that had earlier served to regulate telecommunications system development.[11] This "neoliberal networking drive" would require sweeping changes in telecommunications policy, to be sure. It would also require new divisions of labor among humans and computers, and new technologies of identification better suited to the demands of "network security." In turn, biometric technologies would be envisioned and designed to meet the demand not only for more secure computer networks but also for more intensified, automated forms of surveillance, access control, and identification-at-a-distance in a wide range of settings.

This chapter examines how automated facial recognition and related technologies were envisioned and designed to serve a set of institutional priorities during the period of political-economic neoliberalization in the United States. Neoliberalism has not been an entirely unified program so much as an ad hoc set of governmental experiments that favor privatization, free-market principles, individualism, and "government at a distance" from the state system.[12] It has involved new ways of allocating government among state and non-state actors, a configuration that began to take shape in the United States and other developed nations in the 1960s and 1970s and gathered momentum in the 1980s and 1990s—precisely the period during which computerized forms of facial recognition and expression analysis became a possibility.[13] In the United States, the effort to program computers to identify human faces began in the 1960s in research labs funded by the Department of Defense and intelligence agencies. By the 1990s, new companies were formed to commercialize the technology, searching for markets especially among institutions operating proprietary computer networks (like the finance industry and other business sectors) and large-scale identification systems (like passport agencies, state Department of Motor Vehicle offices, law enforcement, and penal systems). Across these sectors, biometric identification promised to enable what Nikolas Rose has referred to as the "securitization of identity," the intensification of identification practices at a proliferation of sites—a priority that has gone hand in hand with political-economic and governmental neoliberalization.[14]

A close look at the early commercialization of facial recognition and other biometric technologies suggests that the turn to biometric identification at this particular juncture should not be understood as the inevitable result of seemingly inherent inclinations of the nation-state toward sovereign control of populations and territory, fueled by advancements in computer science and visual media technologies. Certainly technological advances and the persistence of ideas about state-centered forms of political power play their part. But the transition to biometric identification must likewise be understood as a response to a set of conflicting demands of both the state and the business system to individualize and to classify, to include and to exclude, to protect and to punish, to monitor and define parameters, and to otherwise govern populations in the face of their radical destabilization under the wrenching neoliberal reforms instituted in the United States and across the globe during the latter part of the twentieth and early twenty-first centuries. Building on existing communication and identification practices that gave pride of place to the face, facial recognition technology promised to play a unique role in this process. Although posing formidable technical and logistical challenges, functioning facial recognition systems promised to build on existing identification infrastructures to refashion face-to-face relations in networked environments, appropriating face-to-face forms of trust and recognition for official identification in mediated contexts. The primary aim of these systems would be to deliver concrete, practical benefits in the form of more accurate, effective, ubiquitous systems of facial identification that operated automatically, in real time, and at a distance. Although they seemed to simply recognize people the way most humans do in their everyday lives—by looking at one another's faces—facial recognition systems in fact promised to enable more effective institutional and administrative forms of identification, social classification, and control.

## *Early Research on Facial Recognition Technology*

Since its beginnings, research into computerized facial recognition in the United States has been a combined public-private venture funded and shaped to a significant extent by military priorities, a fact that makes it far from unique. Some of the earliest research on machine recognition of faces can be traced back to the 1960s at a private company called Panoramic Research Inc. in Palo Alto, California, one of many companies started up in the United States, post-*Sputnik*, to conduct government-funded research in computer science. The work at Panoramic Research was funded largely by the U.S. Department of Defense and various intelligence agencies, and so was unavoidably entrenched in the struggle for Cold War technological superiority. Although not all the computer scientists working in the postwar context were steadfastly dedicated to fulfilling military needs, they nevertheless had to emphasize the applicability of their work to Cold War priorities in order to secure funding from the Defense Advanced Research Projects Agency (DARPA) and its Information Processing Techniques Office (IPTO).[15] Automated facial recognition in particular might eventually help the military identify, at a distance, specific individuals among the enemy ranks, in this way contributing to what Paul Virilio calls a "logistics of military perception."[16]

Aside from the potential military applications of the technology, scientists working on early efforts to simulate face perception in computer programs were not doing so in response to immediate or well-defined social needs. In a manuscript of his dissertation published in 1977, Takeo Kanade speculated that the techniques of picture processing to which his research contributed might lend themselves to "sophisticated applications such as interpretation of biomedical images and X-ray films, measurement of images in nuclear physics, processing of a large volume of pictorial data sent from satellites, etc."[17] The techniques being developed for computer recognition of faces promised to address a set of vaguely defined problems concerning how to automatically process images and handle an expanding volume of visual information in medicine, science, and military intelligence. More broadly, early research and development of computerized facial recognition was part of a general effort to program computers to do what humans could do, or, better, what humans were incapable of doing. For some computer scientists and engineers conducting work on pattern recognition, as Manuel De Landa has noted, "the idea was not to transfer human skills to a machine, but to integrate humans and machines so that the intellectual skills of the former could be amplified by the latter."[18]

This would be no simple undertaking. A look at the early work on machine recognition of faces at Panoramic Research underscores the formidable challenges ahead of computer scientists interested in creating computer programs that could identify faces in images, and the significant amount of human effort that would be required. Woodrow Wilson Bledsoe, one of the cofounders of Panoramic, headed up the research. Bledsoe is now recognized as a pioneer in the field of "automated reasoning" or automatic theorem proving, an arm of early artificial intelligence research. A member of the Army Corps of Engineers during WWII and a devout Mormon, he was a firm believer in incremental scientific advances as opposed to major leaps or paradigm shifts.[19] Drawing on his research into computer recognition of letters, Bledsoe's technique involved manually entering into a computer the positions of feature points in an image, a process known as "feature extraction." A human operator would use a "rand tablet" to extract the coordinates of features such as the corners of the eyes and mouth, the top of the nose, and the hairline or point of a widow's peak.[20] The name of the person in an image was stored in a database along with facial coordinates, and records were organized based on those measurements. The computer was then prompted to identify the name of the closest test image, given a set of distances between facial feature points. Bledsoe's technique was labeled a "hybrid man-machine system" because a human operator was centrally involved in the process of extracting facial coordinates from the images.[21] In addition to relying significantly on human intervention, Bledsoe's man-machine program brought to light the fundamental difficulty that computer programs would have with facial image variability, especially in terms of "head rotation and tilt, lighting intensity and angle, facial expression, aging, etc.," thus introducing the need for techniques of image "normalization," or the manipulation of facial images to correct for angle, lighting, and other differences that confounded the matching process.[22]

Following on the heels of this early research, computer scientists made small advances at programming computers to recognize human faces in images in the early 1970s. A very basic problem that proved difficult was programming a computer to simply locate a face in an image, especially if the image was visually cluttered, if faces were not represented in frontal view, or if faces were occluded with beards, hats, eyeglasses, or other objects. The earliest work to successfully program a computer to confirm the existence or absence of a face in an image, without human operator intervention, was conducted by three Japanese computer scientists and published in the journal *Pattern Recognition* in 1969.[23] Then in 1970, a doctoral student produced a landmark dissertation project at Stanford.[24] His technique enabled the computer to automatically extract the head and body outlines from an image and then locate the eyes, nose, and mouth, using three images of each individual: an image of the body, an image of the background without the body, and a close-up image of the head. In 1973, Takeo Kanade's dissertation at Kyoto University in Japan made another significant contribution, reporting the same results as the Stanford research using only photographs of the face and a new "flexible picture analysis scheme with feedback," consisting of a collection of simple "subroutines," each of which worked on a specific part of the picture.[25] The recognition phase of Kanade's project, which focused on the problem of automated extraction of face features, correctly identified fifteen out of twenty people in collections of forty photographs.[26]

Experiments were not always designed with the goal of eliminating humans entirely from the process of recognizing faces. Instead, researchers had in mind creating synergistic relationships between human brains and computers. In 1971, scientists at Bell Labs worked on a system designed to locate and rank order a set of facial images from a file population based on a verbal description inputted into a computer.[27] In their report, titled "Man-Machine Interaction in Human-Face Identification," they explained their aim "to design algorithms for optimizing the man-machine system so that we can take advantage of both the human's superiority in detecting noteworthy features and the machine's superiority in making decisions based on accurate knowledge of population statistics."[28] The research had the more limited goal of producing a "population-reduction" system, a means of reducing large numbers of images to a quantity more manageable for perceptual and memory capacities of human operators.

By the mid-1970s and early 1980s, more limited goals became the norm in computer face recognition research.[29] The initial successes achieved by pioneers like Bledsoe and Kanade generated optimism among researchers that the problems could be surmounted in short order, but such views soon gave way to more a realistic sense of the limitations of existing computing technology and the need for a better understanding of the "physics" of vision. In short, it became increasingly clear that the general problem of programming computers to identify human faces was more difficult than anyone had initially realized. Significantly, while humans lived in a three-dimensional world, computer vision systems were designed using two-dimensional representations of that world, and that flat field of vision severely handicapped a computer program's capacity to recognize objects. The amount of computing power that was needed to process images also presented a major obstacle. Realistically, computer vision researchers would have to get back to basics, especially to carve out what could be accomplished with measurement techniques versus what required heuristics, problem-solving algorithms that would find practical solutions among alternatives through trial and error at multiple stages of a program.

The formidable challenges to developing automated facial recognition paralleled—and were an instance of—broader challenges to the development of artificial intelligence. As it turned out, the human perceptual capacity to see and recognize human faces was more than a simple matter of "information processing"—either that or computers were going to need substantially more processing power. As De Landa has noted, solving the computer vision problem would require solving the central problems of artificial intelligence:

the first computer to "perceive the world" would have "to be able to learn from its successes and failures, plan problem-solving strategies at many levels of complexity and have a certain amount of 'common sense' to avoid getting bogged down by irrelevant details."[30] But as De Landa has also explained, the failure to develop machines that can see the world the way humans do has not prevented computer vision technology from finding practical applications. As it has been with other forms of computer vision and artificial intelligence, so it is with automated facial recognition: alternative uses of the technology have been envisioned and pursued that do not involve replacing humans entirely with computers, but instead work toward developing new, optimized human-computer divisions of perceptual labor. "Instead of building computers to automate the process of bringing patterns to the surface," writes De Landa, "the surface itself (the computer display) had to become a place where the human ability to detect patterns may be amplified."[31] The idea that computerized forms of facial recognition might be used to amplify the human capacity to recognize faces became a promising possibility, especially as the problem of identification-at-a-distance began to take on a new level of urgency.

## *Securitization of Identity*

Early research on automated facial recognition technology did not take place in a vacuum, and it is important to understand the relationship between this area of computer science and the broader social and political-economic transformations in which it was conceptualized and undertaken. Where a technical explanation would describe this evolving area of research as an outgrowth of specialized problems in computer science, the aim of this book is to examine the social construction of automated facial recognition, and especially to place it in its broader social and historical context. How did research on computer recognition of faces join up with the demand to re-embody "disembodied identities"? Although in some ways an obvious marriage, it was by no means the inevitable result of the internal workings of scientific and technical progress. Nor was it driven exclusively by military priorities or the narrowly defined interests of computer science. By the 1990s, the effort to teach computers to "see" the face began to find prospective social uses and potentially profitable markets in the surveillance and security priorities that underpinned neoliberal strategies of government—a set of approaches to governing designed to be maximally commensurate with privatization and market-driven tenets, involving new ways of allocating the tasks of government among state and non-state actors. Envisioned as a biometric identification technology that could be integrated into existing identification systems to extend their reach and make them function more automatically and effectively, computerized facial recognition promised to facilitate the forms of mass individuation, social differentiation, and intensified security on which neoliberalism depended.

While it is important to understand the immediate political-economic and social context for early uses of facial recognition technology, the application of new photographic technologies to institutionalized identification practices is part of a process that dates back at least to the nineteenth century: the rationalization of identification for the mass individuation of populations. The development of systematic techniques and procedures for mass individuation have been central to processes of bureaucratic rationalization that Max Weber theorized as a core feature of Western capitalist modernization.[32] The rationalization of facial identification was greatly assisted by the invention of photography, especially the technical innovations of the 1880s and 1890s that enabled that mass production of cheap, photomechanical reproductions.[33] The circulation of facial images as indexes of identity occurred within the more general formation of what Craig Robertson has called a "documentary regime of verification," an ad hoc process whereby techniques for administering standardized identification documents and using those documents to verify identities came to replace local practices of trust and face-to-face recognition.[34] The expanding scale of modern societies, and the mobility and anonymity of people within those societies, necessitated the "depersonalization of authority and trust" through recourse to standardized documents, administrative procedures, and archives of "state memory."[35] State passport systems were a case in point. As John Torpey has argued, the emergence of the modern passport system was an effort on the part of state systems to "monopolize the legitimate means of movement," akin to the way states likewise laid claim to the legitimate means of violence, as Weber argued, and to the way capitalists had monopolized the means of production, in Marx's central thesis. While photographs of the face could be found on identification documents as early as the 1880s, it was the First World War that spurred many countries to begin mandating that photographic portraits be fixed to passports. In 1914, the U.S. secretary of state moved to require the inclusion of photographs on all new passports issued; however, the design of the document did not include a space for photographs until 1917.[36] Even then, the U.S. public was reluctant to accept the requirement of photo passports because of the criminal connotations of the mug shot—one indication of the extent to which the bureaucratization of identification practices has been a contested process.[37]

Certainly the preoccupation with improving on bureaucratic identification systems, along with the persistent cultural resistance to new identification technologies, has been with us for some time. But over the last several decades, the sites at which individuals are asked to verify their identities have proliferated, extending their reach well beyond interactions between citizens and state systems. The development of biometrics and machine-readable documents, the expansion of databases that house personal information about individuals, the adoption of new policies for identification systems (like US-VISIT and the U.S. REAL ID Act)—all these developments point to an intensification of the problem of identification in these times. "At the close of the twentieth century," writes Nikolas Rose, "subjects are locked into circuits of control through the multiplication of sites where exercise of freedom requires proof of legitimate identity."[38] Active citizenship is now realized not primarily in voting or participating in an idealized political public sphere, Rose argues, but through consumption, employment, financial transactions, and other practices, virtually all of which require the verification of legitimate identity.[39] What Rose calls the "securitization of identity" is widely viewed as a solution to problems of network security, consumption, labor, crime, welfare, immigration, and more—a means of tying individuals into circuits of inclusion and exclusion, determining their status as legitimate, self-governing citizens or more problematic identities deemed to fall outside the realm of legal rights who can therefore be subjected to a range of repressive strategies. Although the rationalization of identification practices was never exclusively a state project (see Josh Lauer's early history of credit reporting discussed below, for example), more recent developments in identification systems have seen an intensified involvement of non-state actors as both suppliers and users of identification technologies. The respective roles of state and private-sector actors in the development and use of surveillance and identification systems are increasingly difficult to disentangle, with both sectors exhibiting an intensified concern with the securitization of identity, adopting similar approaches and systems, and sharing information about individuals across sectors.

The tightly bound relationship between state agencies and corporate actors in identification system development is itself a manifestation of the neoliberal experiment. David Harvey explains neoliberalism in general terms as the effort to disembed capital from the constraints of "embedded liberalism," a form of political-economic organization that took shape in the United

States and other developed countries after World War II.[40] In order to prevent a return to the conditions that threatened capitalism during the Great Depression, liberal democratic states took active measures in the postwar period to wager a class compromise between capital and labor, to actively regulate industry, and to establish a variety of welfare systems.[41] Embedded liberalism surrounded market processes and corporate activities in "a web of social and political constraints," instituting a regulatory environment that both restrained and shaped economic and industrial strategy.[42] Although offering high rates of economic growth for a time, thanks in large part to expansive U.S. consumption, embedded liberalism began to break down by the end of the 1960s, when "signs of a serious crisis of capital accumulation were everywhere apparent."[43] In response to the threats this crisis posed to their political and economic power, elites moved to institute a set of reforms designed to dismantle the postwar social contract, pushing for "the deregulation of everything from airlines to telecommunications to finance" in order to open up "new zones of untrammeled market freedoms for powerful corporate interests."[44] Neoliberalism, Harvey maintains, has essentially been "a political project to re-establish the conditions for capital accumulation and to restore the power of economic elites."[45]

It is no coincidence, then, that the period of neoliberalization has also been a period of frenzied development in information technologies, the expansion of proprietary computer networks, and concerted attention to devising more sophisticated techniques of surveillance and identification. The response of economic and political elites to the economic crises of the late 1960s was tied in complex ways to processes of computerization and network development. As Dan Schiller has argued, the central role that elites accorded to information and communications as economic stimulants was unprecedented.[46] Corporate capital invested intensively in the development of information and communication technologies (ICTs) and undertook a major overhaul of business processes in order to integrate ICTs, especially to coordinate dispersed locations. As an integral part of the process, telecommunications infrastructure worldwide experienced a top-down overhaul during the 1980s and 1990s to bring it in line with the needs of transnational business operations, and the U.S. policy-making establishment was necessarily onboard, "determined to grant business users the maximum freedom to explore information technology networks as a private matter."[47] Widespread decisions to interoperate computer systems were motivated by the interests of large corporations to spread their reach across the globe and deep into economic, social, and cultural life. While corporate reconstruction around networks occurred economy-wide, the financial sector took a leading role in this process. Banks significantly increased their telecommunications operating expenses, linked up their offices transnationally, and installed 165,000 automated teller machines in the United States by 1998.[48] But banks were not alone, as companies in other sectors "sought to integrate networks into core activities of production, distribution, marketing, and administration."[49]

Computerization thus involved sweeping changes in the structure and policy of telecommunications, and along with these changes came demands for new ways of monitoring and restricting access to that infrastructure. The self-proclaimed proprietors of expanding computer networks saw the need to automatically identify, observe, and record the behaviors of the growing numbers of end users communicating and conducting transactions over those networks. Identification documents, like passports and driver's licenses, which had always had their shortcomings—especially in terms of their inability to accurately and reliably connect bodies to identities—were seen as increasingly inadequate for addressing the problem of "disembodied identities." The banking, credit card, and telecommunications industries were among those social actors expressing an interest in technologies that could give them greater control over transactions and information. Not only was identity verification needed at the point of interface between individuals and organizations, but the intensive private-sector drive to know the consumer meant that each transaction became part of individual records that could be mined and compiled to develop consumer profiles. In addition, employers of all sorts saw the need to monitor and control their employees' access to both computer networks and the physical space of the workplace. In a society where an increasing number of social interactions and economic transactions were occurring via proprietary computer networks, biometric technologies were being posited as a solution to a set of pressing problems, such as verifying the legitimate status of individual users, monitoring their behaviors, and controlling their access to networks, as well as to the spaces and entities over which those networks extend: national territories, transportation systems, financial institutions, spaces of production and consumption, and, of course, information itself, which was taking increasingly commodified forms.

Thus, while computer scientists continued to grapple with the complexity of the facial recognition problem during the 1980s, new markets for other biometric identification technologies began to take shape. Attempts were made to commercialize a number of different types of biometrics that were reaching some degree of maturity, such as optical fingerprinting, hand geometry, retinal identification, and voice and signature recognition. The press took notice of the emerging biometrics industry. In September 1981, the *New York Times* first reported on the new business sector of "biometric access control" devoted to developing and marketing technologies aimed at "Recognizing the Real You."[50] The question of what makes each person unique "is of more than philosophical interest to those in the field of biometric access control," the *Times* explained. Advocates of new biometric technologies claimed that they could digitally read bodies and bind them to identities, thereby securing the valuable spaces of the information economy. In a world where "high school students were able to tap into computer data banks thousands of miles away over the telephone," existing security systems that utilized plastic cards and passwords might not be adequate any longer.[51] While the actual amount of high school hacking at the time is impossible to know, the image of the high school hacker held considerable sway in the public imagination. In the 1983 John Badham film *War Games*, for example, the young protagonist (played by Matthew Broderick), nearly starts World War III when he unknowingly hacks into a central military computer and begins playing the system like a computer game. Such images went far in convincing people of the need for "network security." After all, if high school students could wreak havoc via computer networks, then certainly more malicious actors could do so as well. Technologies that enabled the computer to read the unique physical characteristics of individual bodies seemed the perfect solution to the newly intensified problem of verifying identities (binding those identities to bodies) over networks. As one "biometrics expert" told the *New York Times*, "The computer might now help solve a problem it helped to create."[52]

Electronic banking in particular seemed to necessitate more rigorous and reliable ways of identifying people engaged in mediated transactions, as well as preventing unauthorized users from tapping into computers and manipulating accounts from afar. The finance industry's pursuit of automated forms of identification and network security has been a major driving force in the institutionalization of biometric technologies, and so deserves closer attention. A 1982 article in the *American Banker* discussed transformations in the retail banking industry and corresponding needs for new and improved identification systems:

> The traditional scene of a loyal bank customer being greeted familiarly by a bank teller or branch officer carries more than a message of good relations. In that scene is also a bank's most secure method of retail customer identification: direct person-to-person recognition.

> The banking business is changing. Teller-based transactions are being slowly replaced by self-service automated teller machines, offerings of home banking are beginning, and geographic expansion by banks is in many cases making obsolete the "neighborhood" concept of customer identification based on personal recognition.
>
> As part of this trend to convenience banking is a basic depersonalization in the financial service industry—and the growing concern of how to better verify the customer at the point of transaction.[53]

Identity verification problems became particularly salient in the banking industry with the growth of electronic funds transfer (EFT). "Increasing losses to banks from transactions in EFT situations," reported the *American Banker*, "should slowly permit the introduction of dynamic signature verification and other methods."[54] Banks became early adopters of biometric technologies, testing systems for controlling employee access and also envisioning how the technologies might be extended for automatically verifying the identity of banking customers.[55]

As the *American Banker* explained it, trends in the financial service industry "depersonalized" relationships that individual customers once had with banks (and bank workers in particular), necessitating new ways of verifying customer identities that compensated for a lack of consistent, interpersonal, face-to-face relations. But the nostalgic image of trusting personal relationships between bank tellers and bank customers presented in this narrative of transition glossed the reality of banking practices and the significant changes occurring in the finance industry. While some banking customers may have known their local bank tellers, the "neighborhood concept of customer identification" was hardly the predominant mode of banking, nor were face-to-face financial relations ever characterized by certainty, security, and trust. The idea that "direct person-to-person recognition" represented banks' "most secure method" of retail customer identification is at odds with the historical record, given that "depersonalization" was part of the very process of formation of modern financial institutions. As Josh Lauer has shown, the credit reporting industry began in the United States in the 1840s in order "to facilitate safe business relationships in a world increasingly inhabited by strangers," and one of the most consequential effects of the rise of a credit reporting apparatus was "the invention of disembodied financial identity."[56] Coming more than a hundred years later, digital biometric identification would not be a matter of correcting a brand new problem of trust just now being introduced by the computerization of banking.

But nor was the turn to biometrics in the realm of finance simply a matter of upgrading identification systems to meet the demands of naturally evolving industry infrastructures and practices. The problem of "disembodied financial identities" intensified significantly along with the reorganization of banking around networks and the global expansion of the financial system. Banking, and finance more generally, not only was growing by leaps and bounds in the 1980s and 1990s, it was experiencing massive restructuring as a result of the financialization of the overall economy. As Harvey explains, "neoliberalism has meant, in short, the financialization of everything," a deepening hold of finance over other areas of the economy, over the state apparatus, and over everyday life.[57] Economic power shifted from production to finance, making the integrity of the financial system the central concern of neoliberal states.[58] And while proponents of this "new economy" touted its wealth-generating benefits, the "main substantive achievement of neoliberalization" was "to *redistribute*, rather than to *generate*, wealth and income."[59] Beginning in the 1970s and intensifying in the deregulatory environment of the 1980s, the financial sector was restructured in ways that virtually guaranteed endemic, legalized theft, what Harvey calls "accumulation by dispossession."[60] It involved particularly aggressive and complex strategies of economic restructuring:

> The strong wave of financialization that set in after 1980 has been marked by its speculative and predatory style. . . . Deregulation allowed the financial system to become one of the main centres of redistributive activity through speculation, predation, fraud, and thievery. Stock promotions, ponzi schemes, structured asset destruction through inflation, asset-stripping through mergers and acquisitions, the promotion of levels of debt incumbency that reduced whole populations, even in the advanced capitalist countries, to debt peonage, to say nothing of corporate fraud, dispossession of assets . . . by credit and stock manipulations—all of these became central features of the capitalist financial system.[61]

The securitization of financial identities and transactions was essential to the explosive growth of the banking network infrastructure that underpinned the financialization of the global economy. But while the securitization of electronic transactions was necessary for consumer confidence in new banking practices, biometric technologies were not going to protect banking customers from the real threats to their financial security posed by the political-economic restructuring of the finance industry. The fact that financial identities were increasingly "disembodied" in all likelihood furthered the aim of making "accumulation by dispossession" seem less like legalized theft and more like a "free market" finance system at work.

While credit reporting and its "disembodied financial identities" had been around for some time, they took on new life in the 1980s not only as a result of the consolidation of banks and the expansion of banking network infrastructures, but also as a result of the dramatic growth of the credit industry. One of the most devastating and radical means of accumulation by dispossession,[62] the credit system has likewise depended for its existence on secure forms of customer identification and surveillance. Credit cards represent a case in point. In the United States, relaxed regulations in the 1980s allowed credit card companies to increase their interest rates, making the industry much more profitable. Credit card companies issued more cards, raised credit limits, and saw credit card spending increase more than fivefold between 1980 and 1990.[63] Many more merchants began accepting credit cards, and a whole array of clubs, charities, professional associations, and other nonfinancial institutions began issuing affinity or "co-branded" cards with their own logos.[64]

Increasing credit card usage and the expanding transaction infrastructure exacerbated the problem of cardholder identity verification that had existed since department stores began issuing cards in the early decades of the twentieth century. Whereas paying cash never required an identity check (unless there were restrictions on the product or service being purchased), credit card transactions were more problematic since cards could easily be lost or stolen or obtained in other fraudulent ways. In his analysis of the BankAmericard system in the early 1970s, the sociologist James Rule commented on the problem of verifying the identity of individuals in order to authorize transactions:

> One element of the contact between system and clientele . . . is the ease of establishing positive identification of clients. . . . [The BankAmericard] system has nothing like the virtually foolproof tool of fingerprinting which the police can often rely on. Indeed, one category of deviance which poses a most serious threat to the system is the fraudulent use of cards—something which would be quite impossible if means of identifying card users were truly effective.[65]

By the mid-1980s the industry's losses to credit card fraud rose to hundreds of millions of dollars per year, so it was unsurprising to find companies like MasterCard investigating the practicality of using the tools that police rely on—the possibility of integrating fingerprinting into point-of-sale terminals, as the *New York Times* reported in 1985.[66] Developers of biometrics viewed the problem of effective cardholder identity verification as a problem that their technologies were uniquely designed to address—fingerprinting would no longer be just for criminals. The problem of fraud, endemic to the expanding credit card market (some might say "racket"), enabled biometric companies to position their products as consumer protection technologies. But what the consumer protection claim elided was the fact that widespread credit theft was made possible by the aggressive approach credit card companies took to issuing accounts, spreading their reach, and making it as easy as possible for cardholders to build up credit card debt. As more people became victims of credit card theft and fraud, and as the media took notice and spread the word, the idea of protecting one's accounts with bodily identification technologies probably seemed to many a reasonable and even necessary solution. Whether consumers would in fact accept biometrics was a major concern to credit card companies. Their comments to the press suggested that that they would not move forward with point-of-sale biometric identification systems unless they could be sure that cardholders felt comfortable with the technologies—that is, unless credit card companies knew consumers would continue to use their cards freely, even if it required pressing their fingertips to a plate of glass with each purchase.

The growing replacement of cash with credit cards for consumer purchases not only had the effect of creating conditions in which each consumer transaction became a potential identity check; using credit cards for payment also generated detailed records of each consumer's purchases. These records documented what people bought as well as what "they read, how they played and drank, where they vacationed and where they lived," along with many other things about their everyday lives.[67] The transaction data being gathered by credit card companies would become a valuable commodity in itself, to be processed and analyzed along with data from other sources in increasingly complex and automatic ways to classify individuals into distinct categories of value and risk. As Joseph Turow has documented, competition among market research firms led to a "market research arms race" in the 1980s.[68] The increasing sophistication of statistical tools combined with the use of computers and large databases of transaction-generated data allowed market researchers "to find clusters of relationships among demographic, attitudinal, behavioral, and geographical features of a population that marketers hadn't noticed before."[69] This intensified market research apparatus, armed with new scientific and technical tools, was clearly engaged in a process that

Ian Hacking has called "making up people," in this case in order to more thoroughly rationalize consumption: matching consumer demand to cycles and patterns of production, with the aim of harmonizing consumption with productive capacity, and especially increasing consumption to levels well beyond necessity.[70] Although biometric technologies were designed to individuate rather than to aggregate consumers into categories, a reliable means of identification was central to this rationalization process. Identifying individuals with certainty was obviously a key part of determining their appropriate consumer classifications. Biometric technologies would be especially useful for customer relationship management (CRM)—the targeted tracking of customers in order to give preferential treatment to those of most value—and eventually even new forms of customized advertising (like the futuristic billboards targeting people by name in Steven Spielberg's 2002 science fiction film, *Minority Report).*

The banking, credit card, and consumer research industries had unmistakable roles to play in generating early demand for biometric identification, but they were not the only businesses considering the adoption of biometric systems in the 1980s, nor were potential uses limited to customer identification. Biometric systems would enable the identification and tracking of individuals not only in their roles as consumers but also as workers, promising to provide tools for more effective forms of labor control. Automated identification over networks would be especially useful for individualizing labor, another project that was central to the neoliberal experiment. Beginning in the late 1980s, a range of business enterprises as well as nonprofit and government employers became interested in integrating biometric technologies into personnel time management processes.[71] In addition to preventing workers from punching one another's time cards, biometrics offered the possibility of more meticulous monitoring of the employee's workday. Biometric systems could be used to assign individuals specific authorizations for what they were permitted to do and where they were permitted to go in the workplace, and keeping diligent automatic tracking reports of precisely how and where employees spent their time. As a Booz Allen Hamilton associate remarked in a 1990 issue of *Computerworld*, "Biometrics offers something other systems don't . . . [they] can prove irrefutably that John Doe was the one who came through this door at 11 last night."[72] This comment not only captured the technology's promise to intensify the surveillance of workers, but also the powerful claim to truth made in the name of biometrics: that biometric technologies of all types could "prove irrefutably" the identity of individuals, along with their presence or absence at specific places and times.

Early experimentation with biometric systems for personnel time management and the monitoring of workers' movements and activities was obviously viewed by early adopters as a means of addressing specific organizational and business needs, but interest in these technologies should also be understood as a response to the broader political-economic transformations of the period and their impact on work and employment. The introduction of biometric technologies into the workplace, however fragmentary and sporadic, represented one of a wide array of managerial and technological efforts aimed at instituting new forms of labor control better suited to the post-Fordist regime of flexible accumulation.[73] Biometric identification was especially suited to the individualization of labor, the process by which the contribution of labor to production would be "defined specifically for each worker, and for each of his/her contributions, either under the form of self-employment or under individually contracted, largely unregulated, salaried labor."[74] The ability to precisely identify and individuate workers and associate their identities with the specific work they performed would enable the continuous optimization of work arrangements and divisions of labor. Biometric systems for personnel time management promised to measure the temporal dimension of labor, as well as track the movement of specific bodies through space, both the physical space of workplace facilities and the metaphorical space of information networks. In short, the management of time and space within specific work sites and across dispersed locations required new technologies of labor control, especially technologies for automatically binding identities to bodies over information networks.

There are important connections among the various uses of biometric technologies for network security, consumer tracking, and labor control—namely, each of these uses stemmed from demand on the part of the business sector for more intensified network security and monitoring methods. Since the 1970s, the banking, credit card, market research, and other industries have made increasing demands for new technologies that would allow them to control access to networks and information, and to gather as much data as possible on individuals in order to more effectively rationalize consumption and labor. As businesses built out proprietary computer networks and moved to absorb new realms of information into the market system, network security and control over information became paramount. As the quantity of transaction data mounted, new techniques were needed to make effective use of the avalanche of data. Biometric technologies promised to keep specific information bound to specific bodies, and to control access to networks and information while at the same time helping to build more reliable and complete profiles of individuals. And although it was not necessarily the aim of any particular application, biometric systems had the potential to bind identities to bodies across different spheres of social and economic life, thereby rendering more data-intensive representations of people, making their lives more visible. This effort would of course require considerable investment of resources for standards development and systems integration. In fact, these have been two areas of special concern to the biometrics industry and its customers. Industry players have worked hard to overcome these obstacles to full-scale biometric deployment, given the level of control over networks, information, and people that standardized, integrated systems promise to afford them. As Robins and Webster have argued, new ICTs have been enlisted in order to apply principles of scientific management beyond the workplace. "Cybernetic" forms of capitalism require a new regime of social mobilization characterized by "a heightened capability to routinely monitor labour processes by virtue of access to and control over ICT networks."[75] It would be difficult, if not impossible, to achieve this new regime of social mobilization without technologies for automatically fixing identities to bodies across information networks.

## *Why the Face?*

Biometric technologies promised to bind identities to bodies over networks and facilitate the securitization of identity in finance, market research, the workplace, and other settings. But what would be the merits of facial recognition technology relative to other types of biometrics? Why the sustained investment in the exceedingly complex effort to program computers to recognize faces and distinguish them from one another, especially since there was no guarantee that this effort would result in viable technical systems?

In some ways, automated facial recognition represented simply one technique among a range of similar, interchangeable possibilities, one that presented significantly greater technical challenges and required more intensive investment in research and development. For certain identification systems, the use of optical fingerprinting, hand geometry, or iris or voice recognition might be more workable options. Each new commercially available biometric technology introduced in the 1980s and 1990s had its relative merits and drawbacks as a means of re-embodying disembodied identities and securitizing identification systems. Fingerprint identification was in some ways the most robust and reliable, and had the advantage of being around for over a hundred years.[76] But it also had the decided disadvantage of having criminal connotations, a problem that was seen as an obstacle to "user acceptance." It would have to shed its association with criminality in order to be accepted as a more widespread means of automatically binding identities to bodies. Although the term "biometric" was not typically used in reference to fingerprinting before the 1980s (rather, the term was used in reference to the statistical study of biology since the late nineteenth century), it was during the eighties that fingerprinting began to be conceived as a new high-tech form of identification, thanks in large part to advances in optical scanning technology that made "live scans" possible and brought fingerprinting out of the print era and into the digital age. As optical scanning began to replace the ink-print form of fingerprinting in the United States, developers took the opportunity to redefine fingerprinting as "finger imaging," a cleaner and ostensibly more sophisticated and innocuous technology than that associated with criminal identification.[77]

Less associated with criminal identification, hand geometry biometrics found some early physical access control applications. But hand geometry had the major drawback of being the least unique of biometrics. Since hand measurements were not unique to each individual, such systems would more likely generate "false positives," a problem that limited hand geometry biometrics to identity verification applications, where the system compares hand measurements to already enrolled data in order to determine whether a person is who she claims to be. Hand geometry biometrics would not be suitable for the identification of unknown individuals using one-to-many searches for matching hand measurements.

Along with their efforts to target body parts unique to each individual and to disassociate biometric technologies from criminal applications, developers saw it as especially important to design systems that would be as noninvasive as possible. The aim was to develop forms of biometric capture that would seem to follow naturally from existing identification practices and not make people uncomfortable. Retinal scanning, for example, was considered to be especially unnerving for end users. Developed exclusively by a company called EyeDentify, the technology measured the unique pattern of blood vessels in the eye (different than iris recognition, which was not developed until 1992).[78] It was difficult to use, requiring careful attention to the alignment of the eye in order to capture a usable image, and one of the impediments to its widespread diffusion, according to the developer, was "the ID subject's unfounded fear" that it could harm the eyes.[79]

The seemingly less intrusive technology of voice recognition, or voiceprint analysis, aimed to verify identity by authenticating the voice of the speaker. Although analog versions of voiceprint analysis had been around for a long time, new automated, digital versions of voice recognition were of special interest to the banking industry and other sectors that were moving toward telephone transactions.[80] But developers faced the difficult problem of the medium: a person's voice could sound very different depending on the recording device or mode of transmission. Also, like hand geometry, it was best suited to verifying the identities of known voices rather than identifying unknown speakers.

Although still in a very nascent stage of development in the 1980s and 1990s, automated facial recognition promised certain relative advantages over other types of biometrics. Proponents suggested that end users would find the technology less obtrusive, since it simply recognized people the way humans do, by looking at their faces.[81] Since the signals people use to make determinations about the trustworthiness of others in immediate face-to-face contexts involve, among other things, their facial expressions, gaze, and appearance, it seemed logical that computational techniques of facial analysis would be envisioned as ideally suited to establishing trust in mediated, networked environments.[82] But facial recognition technology was not designed to analyze the expressions or appearance of the face itself for signs of trustworthiness, instead treating the face as a blank somatic surface, a mere index of an individual identity much like a fingerprint. Its capacity to establish conditions of trust depended on its integration into complex bureaucratic systems rather than its ability to read faces themselves for signs of trustworthiness. Of course, facial images had been used in precisely this manner at least since the late nineteenth century: as markers of individual identities pasted to documents and integrated within larger bureaucratic ensembles. In fact, it is only possible to conceive of digital photographs as source material for automated identity verification because the awe that photographic resemblances originally inspired has long since been replaced by the banality of the portrait, fixed to an identification document. It was an earlier "documentary regime of verification" that incorporated the iconic connection between portraits and their subjects into more standardized bureaucratic procedures for the authentication of specific identities. The connection that documentary forms of identification established between standardized facial images and unique, official identities has helped to give facial recognition technology a special advantage relative to other types of biometric identification.

Put simply, the integration of facial recognition technology into identification systems would have the practical advantage of being able to build on the existing bureaucratic norms and practices of facial identification.[83] The fact that billions of photographed faces already circulated on documents and resided in archives made the application of facial recognition technology for biometric identification seem self-evident. With nearly everyone in the modern world accustomed to carrying a photo ID and presenting it on demand in a wide range of contexts, automated facial recognition systems could be positioned as continuous with document-based identification. In addition to building on existing customs and practices, facial recognition systems promised to solve many of the problems associated with a documentary regime of verification by taking advantage of the memory capacity of computers, compensating for the perceptual fallibility of humans, and binding identity definitively to the body. In addition, if facial recognition technology could be successfully integrated with video surveillance, it would allow for identification "at a distance," without people directly interfacing with identification systems (that is, without their knowledge or cooperation).[84] If the technology could be integrated with video surveillance systems, it would automate some of the perceptual labor of monitoring those systems, helping to increase their scale and reach. Ultimately, the aim of facial recognition technology has been to automatically articulate digitized visual images of the face to documentary information about individuals, and it is the prospect of more thoroughly and effectively integrating visual and documentary modes of surveillance that has made it so potentially innovative and powerful.

## *Commercializing Facial Recognition Technology*

While efforts to commercialize other types of biometrics were underway in the 1980s, it was the post–Cold War decade of the 1990s that saw the early emergence of prototype applications of facial recognition technology beyond the walls of computer science laboratories. Research interest in facial recognition technology increased significantly in the 1990s, along with efforts to develop commercial systems and experiment with real-world applications.[85] Computer scientists made advances in computational methods for locating the face in an image, segmenting it from background clutter, and automatically extracting features. Increasing amounts of computing power facilitated faster techniques and accommodated larger, higher-quality images, and in greater quantities. By the mid-1990s, a number of scientists working in academic research labs had remade themselves as entrepreneurs, hoping to ride the IT economic boom and capitalize on the needs of businesses and other organizations to control access to networks and monitor individual constituents, be they customers, employees, citizens, or criminal suspects. But the technology still faced formidable obstacles to viability, and questions remained about whether the various technical approaches being taken to the problem were ready for the ambitious transition to real-world applications. Migrating facial recognition technology from computer labs to actual surveillance and identification systems would not be simply a technical matter, nor would it be the work solely of computer scientists. Instead, it would require the investment and cooperation of a host of social actors, some with competing or incommensurate interests, including companies trying to capitalize on the technology, organizations looking for cost-efficient ways to secure access to their networks and monitor their workers and customers, and government agencies charged with various responsibilities related to the technology, like subsidizing research and development, testing technical systems, negotiating technology standards, establishing a conducive policy environment, and procuring biometric technologies in order to securitize state identification systems.

In addition, proponents of facial recognition technology would have to make a compelling case to stakeholders that the machine recognition of faces was both necessary and *possible*.[86] Whether machines could be made to *accurately* identify human faces became an important question for developers to answer affirmatively, and precisely how much accuracy would be necessary required some negotiation to define. Despite the seemingly linear progression to computer-assisted identification, the effort to automate the process of facial identification—using the iconicity of facial images as a means of establishing their indexicality—in fact reintroduced the question of precisely how a photograph represents the face, and what exactly constitutes an "accurate" facial likeness. In short, the social construction of the *accuracy* of automated facial recognition became central to its development. Developers began to achieve respectable rates of accuracy in laboratory settings, providing momentum for ongoing research, but laboratory results could not be extrapolated to unconstrained, real-world contexts. This meant that in order to make a case for its viability, inventors of facial recognition technology would eventually need to make use of the real world as their laboratory.[87]

The social construction of the technology's accuracy got some much-needed assistance from U.S. federal agencies in the 1990s. Since the earliest work in the 1960s, the bulk of funding for research in the United States came from military agencies, and by the early 1990s it was time to begin to more systematically evaluate the fruits of those federally financed efforts. In 1993, the Defense Department's Counterdrug Technology Development Program, the Defense Advanced Research Projects Agency (DARPA), and the Army Research Laboratory (ARL) organized the Face Recognition Technology (FERET) program. The purpose of FERET was "to develop automatic face recognition capabilities" that could be employed "to assist security, intelligence, and law enforcement personnel in the performance of their duties."[88] In short, the tests were designed to evaluate the viability of the technology for use by government agencies. The results of the FERET tests generally showed that while some approaches did better than others in the various tests performed, automated facial recognition appeared to be emerging as a viable technology, at least with respect to still images taken under controlled conditions. According to the report, "the overall performance for face recognition algorithms had reached a level of maturity" that warranted their transition "to a real-time experimental/demonstration system."[89] The main purposes of an experimental system, according to the FERET report, would be to develop large-scale performance statistics, demonstrate the capabilities of the system to potential end users, and "identify weaknesses that cannot be determined in laboratory development."[90]

If government testing had the direct result of providing technology benchmarks and identifying problem areas, it also lent legitimacy to the technology, helping to convince stakeholders of its viability for real-world identification systems. Government testing not only encouraged further research, it also inspired developers to begin making moves toward claiming the real world as their laboratory, introducing prototypes into actual surveillance and identification systems. The emergence of new companies dedicated to developing and marketing facial recognition technology happened around the same time as the initial FERET tests. While a technical explanation would point to the progress of the research as the main reason why decisions were made to move forward with commercialization, the close proximity of the FERET tests with the technology's commercial debut was as strategic as it was technically expedient. Entrepreneurs interested in commercializing the technology were able to capitalize on the publicly funded FERET tests, using the favorable results as a springboard to jump-start new companies. Between 1994 and 1996, three U.S. companies emerged as the most promising and visible vendors, vying for venture capital, federal funding, and early-adopting customers: Visionics Corporation, Miros Inc., and Viisage Technology. In 1994, Joseph Atick and two other researchers from the Computational Neuroscience Laboratory at Rockefeller University formed Visionics Corporation, using the relatively strong performance of their technology in the FERET tests as a selling point. Michael Kuperstein, a former scientist at MIT, also started his company, Miros Inc., in 1994. Established two years later, in 1996, Viisage Technology was spun off from the defense contractor Lau Technologies in order to raise capital to finance the expansion of its "high-tech identification card business."[91] Viisage purchased the rights to the facial recognition algorithm developed by Alex Pentland and his colleagues at MIT.[92]

These new companies took up the charge to develop real-time experimental systems and to make a strong case for the viability of the technology through promotional strategies and product boosterism. One approach they adopted was to boast about the superb scientific credentials of corporate executives.[93] Industry discourse favored the Bill Gates model of entrepreneurialism, combining romanticized notions of technological genius with savvy business sense and a conservative-libertarian philosophy. This ideal seemed nowhere more clearly embodied than in figures like Joseph Atick and Michael Kuperstein. A *Newsweek* profile of Kuperstein made a point of noting that he had coauthored a book on neural networks in 1988 and developed a robot driven by a neural network that had the distinction of being written up on the front page of the *New York Times*.[94] Promotional material emphasized that Kuperstein was a former scientist at MIT and an expert in neural networks, also known as "connectionist" computational models, or "networks of simple interconnected processing units" that presumably simulate brain function by learning for themselves.[95] Of course, the very use of the term "neural network" in reference to a computer program implied that computers could now do what the human brain could do, at the very least renewing the lofty promises of artificial intelligence, and Kuperstein's expertise in this area clearly positioned him as on the technological vanguard of computer science. The press profile of Joseph Atick likewise portrayed him as supremely intelligent, a child prodigy who wrote a six-hundred-page physics textbook at age sixteen and earned a PhD in physics from Stanford by the time he was twenty-one.[96] Clearly accomplished scientists, men like Kuperstein and Atick seemed to possess the entrepreneurial spirit needed to make automated facial recognition not just technically viable but also useful and even profitable, transforming it from a set of laboratory experiments into a new, more sophisticated and expensive identification technology.

The effort to migrate automated facial recognition from computer science laboratories to real-world identification systems was a public-private venture, with both state and non-state actors working together to define the technology's viability and use value. If academic computer scientists were primarily interested in creating intelligent machines and more sophisticated image processing techniques in their research labs, computer-scientists-turned-entrepreneurs were interested in using real-world settings as their labs, attempting to build profitable IT companies by responding to the surveillance and security needs of large-scale business and government users. As with other biometrics, business demand for more effective forms of network security, consumer tracking, and labor control provided a promising path to institutionalization for commercial facial recognition systems, especially as computing technology became better able to manage a growing volume of visual images. As the process of upgrading government ID systems was outsourced to private companies, and as the business sector made demands for more secure forms of government-issued identification, the respective roles of state versus private-sector actors in both the supply and demand for biometric technologies became increasingly difficult to disentangle.

### *State Support for Securitized Identity: The Driver's License*

While the private sector has played a central role in biometric system development on both the supply and demand sides, this does not negate the equally important roles played by a variety of state agencies on the federal and state levels. And while the armed forces have played a key part in research and development, state involvement has not been limited to military investment. Civilian agencies have also had a hand in subsidizing research and development, sponsoring tests of vendor systems, negotiating technology standards, and establishing a policy environment conducive for biometric system implementation. In addition, a range of civilian government agencies have furnished demand for the technologies and have done the expensive and labor-intensive work of experimenting with unproven products, especially agencies operating large-scale identification systems, from the State and Justice Departments to individual states' Department of Motor Vehicles offices (DMVs) and welfare agencies.[97] Facial recognition companies have invested considerable effort cultivating public-sector customers, according to Viisage CEO Bob Hughes, because "that's where the money is and the faces are."[98] Working with existing archives of facial images compiled by state agencies, facial recognition systems promise to securitize systems of driver's license administration, passport and visa issuance, border control, airport screening, criminal identification, and other large-scale, public-sector identification systems.

In the United States, driver's license administration in particular has been a special site of experimentation with facial recognition technology. State

DMV offices have served as important experimental test sites, providing real-world laboratories for the development of large-scale systems. In the 1990s, driver's license administration began a major overhaul to incorporate new technologies and systems that would enable individual states to better manage driver licensing and facilitate information sharing among state DMV offices. A recognized problem instigating these major overhauls was the ability of individuals to rather easily obtain multiple licenses, especially in different states, a problem intensified by the fact that the documents were used as a form of official identification across institutional settings. Not surprisingly, facial recognition technology promised to provide a solution to the problem of driver's license fraud, a form of identity fraud. On the heels of the FERET tests, DMV offices began experimenting with automated facial recognition as a means of preventing individuals from obtaining multiple licenses under different names, beginning with the states of West Virginia and New Mexico in 1997.[99] As DMVs replaced laminated film with digital ID photographs, storing the facial images in databases became a viable option, although of course requiring extensive—and expensive—system upgrades. With the necessary computing muscle behind it, facial recognition technology promised to help create more secure driver's license processing by enabling DMV officials to search new photographs of faces against the DMV's database of existing facial images, and DMV databases could eventually be linked together (although DMV system integration would be no simple matter and represents an ongoing challenge).

If the modern passport system emerged as an effort on the part of states to "monopolize the legitimate means of movement" across national borders, as Torpey has argued, then the driver's licensing system has served as an especially apt form of regulating mobility within the territorial boundaries of the nation-state.[100] But driver's licensing has evolved into much more than a system for authorizing individuals to operate motor vehicles. The driver's license has become a de facto all-purpose form of identification, used for a wide range of identity verification purposes beyond permitting individuals to drive. The documents are now used by millions of people to open bank and credit card accounts, cash checks, purchase restricted goods, authenticate their legal status to employers, and otherwise verify their "true identities" in encounters with both state and non-state institutions. The driver's licensing system has been at the center of the securitization of identity and one of the early sites for the biometric registration of citizens in the United States, and the demand for the modernization of state driver's license systems has not come exclusively from agencies charged with governing the roadways. As one of the major identification systems across all sectors, the driver's license has to meet the security demands of industry as well as law enforcement.

Although the initial rationale for integrating facial recognition into driver's license systems has been to prevent applicants from obtaining multiple licenses with different aliases, it is more importantly a critical first step toward widespread institutionalization of the technology. DMV offices represent one of the first major markets for facial recognition systems, and one of the first places significant numbers of people are interfacing with the technology, regardless of whether they are aware of it. Because the driver's license has become a baseline form of official identification, making automated facial recognition a key component of both document issuance and the document itself provides an important route toward more extensive diffusion of the technology. If automated facial identification could be made to work effectively in large-scale DMV systems, and especially if it becomes a normal part of the visit to the DMV, deploying it in a wide range of other settings where identity checks are required becomes a more viable possibility. In short, in the United States, state DMV offices are on the front lines in the securitization of identity and represent key sites for institutionalizing facial recognition technology.

## *Repressive State Apparatuses*

While many U.S. citizens have had their first encounter with facial recognition and other biometrics at the DMV, state agencies have also begun implementing the technologies for a variety of more repressive apparatuses, including criminal identification, welfare administration, and immigration control systems. And just as biometric driver's licenses serve industry demands as much as law enforcement priorities, early efforts to integrate facial recognition and other biometrics into the penal-welfare-immigration-control complex should not be understood as entirely distinct from commercial and workplace uses. Instead, these applications are similarly tied to the neoliberal experiment.

Companies marketing new optical fingerprinting technologies have viewed police departments and prisons as a lucrative market for their products. The police were also identified early on as a major potential market for facial recognition systems. The ever-increasing quantities of photographs compiled by law enforcement continued to present major challenges for the management of criminal identification systems, a problem that dates back to the early application of photography for this purpose in the nineteenth century—in simplest terms, the tendency of photographs to accumulate in disorderly piles, making it difficult to retrieve the identities of criminal recidivists. The persistent need of the police for more effective means of organizing photos and identifying faces explains why the U.S. National Institute of Justice provided nearly as much funding for research and development of the technology as the Department of Defense in the 1990s.[101] Facial recognition and other biometrics promised to provide the police with effective new techniques for both pre-emptive crime-control strategies and post-crime criminal investigations.

Law enforcement agencies also began to adopt facial recognition and other biometrics in the 1990s as tools for managing the exploding prison population. Thanks in large part to the "war on drugs," the U.S. prison population began a steep upward climb around 1980, reaching an unprecedented 1.2 million prisoners by 1991, or 482 incarcerated individuals per 100,000 inhabitants.[102] By February 2000, the U.S. prison population reached an astounding 2 million.[103] Prisons were among the early adopters of biometrics, employed to bind inmates' identities to their bodies in order to better manage and control their growing numbers. One example was the Pinellas County Sheriff's Office in Florida, which adopted an inmate booking system equipped with the Viisage facial recognition system in the summer of 2001. At an initial cost of $2.7 million, the system was designed to capture facial images in order "to meet the operational needs of law enforcement agencies throughout Pinellas County [and] the state of Florida," mainly "to assist law enforcement in identifying suspects and the corrections staff in the sheriff's office in identifying inmates at the time of booking or release."[104] The state of Minnesota had begun implementing a similar mug shot booking system two years earlier, using Visionics FaceIt technology. Called "CriMNet," Minnesota's system would "allow police, judges and court officers to track and share information throughout the state, and increase the access to criminal identification information across jurisdictions."[105] The implementation of networked identification systems, equipped with facial recognition and other types of biometrics, would provide a means of making criminal identities more accessible and distributable.

Intensified interest in new technologies designed to enhance criminal identification systems and circulate criminal identities over networks became part of an overall police strategy of treating disenfranchisement and social disorder as problems best dealt with through technical means. But what this neoliberal crime-fighting strategy elided was the inescapable fact that the crime problem itself was tied to growing economic disparities fostered by a related set of neoliberal policy priorities.[106] Unless one subscribes to the belief that criminal behavior is innate and that the biologically criminal population was reproducing itself in greater numbers, the exploding prison population in the United States over the period of neoliberalization is evidence enough that "incarceration became a key strategy to deal with problems arising among discarded workers and marginalized populations."[107] The "prison-industrial complex" became "a thriving sector (alongside personal security services) in the U.S. economy."[108] The need for more secure and sophisticated criminal identification systems has been part and parcel of what Loïc Wacquant calls the "penalization of poverty"—a set of efforts "designed to manage the effects of neoliberal policies at the lower end of the social structure of advanced societies."[109] The dismantling of social welfare programs, the deregulation of industries, the transformation of finance along with accompanying upward wealth redistribution and accumulation by dispossession, and the destabilization and individualization of wage labor—these neoliberal strategies have gone hand in hand with "the return of an old-style punitive state" that adopts mass imprisonment as an approach to governing the criminalized poor.[110] In the United States, the criminalized poor are funneled into what Jonathan Simon calls the "waste management prison" system, a form of incarceration that has largely replaced the rehabilitation-oriented system of the New Deal period. The waste management prison "lacks an internal regime, whether based on penitence, labor, or therapy (or something else), and increasingly relies on technological controls on movement and the violent repression of resistance."[111] Biometric systems were among an array of technologies designed to track the disenfranchised criminal class as they moved within and outside of prisons, monitoring their movements and activities as necessary in order to manage their risk to society.[112]

Not surprisingly, experimentation with biometric technologies on the part of state agencies was also tied to the "welfare reform" agenda of the 1990s. In 1995, the *Wall Street Journal* reported that biometrics systems, "designed to reduce fraud and improve efficiency, offer a relatively painless way to reduce welfare costs at a time when President Clinton and Congress are haggling over how much to cut entitlement programs."[113] New York was among the first states to implement biometrics for entitlement programs, beginning with two rural counties in 1992. By 1995, the state of New York had spent $4.5 million on its biometric system for welfare recipients, and anticipated spending another $8.5 million.[114] The 1996 Personal Responsibility and Work Opportunity Reconciliation Act recommended that individual states adopt appropriate technologies for limiting fraud in the welfare systems. Although the act did not mention biometric technologies specifically, by 1998 eight states had implemented or started procurement of biometric fingerprint systems for identifying welfare recipients. States also conducted trials of facial recognition systems.[115] The questionable assumption that welfare fraud was a rampant problem, with untold numbers of "double-dippers" and ineligible recipients cheating the system, prompted many states to spend millions of taxpayer dollars adopting new commercial biometric identification systems.

While biometric companies and welfare agencies alike made claims about the millions of dollars in savings that resulted from new ID systems, one would be hard-pressed to find the solid supporting evidence of such cost-benefit analyses. Although they may or may not have paid for themselves, new biometric welfare ID systems no doubt effectively induced in recipients a sense of having their assistance—and their lives—closely monitored. Where promotional rhetoric advocating the integration of biometrics with driver's licenses insisted on the benefits for applicants, pushing the "user convenience" angle, the discourse of welfare biometrics took a decidedly different tone. In partnering with Polaroid to integrate FaceIt into "secure identification product offerings" for DMVs, for example, Visionics promised that the technology would make the process "convenient and non-invasive for the applicant."[116] "In the future," according to the company, facial recognition systems could "allow drivers to renew their licenses at an unattended kiosk in local supermarkets or post offices."[117] In contrast, the discourse of welfare policy showed little concern for whether welfare recipients found biometric identification more "convenient." Nor were there promises that the technology would support kiosks at local grocery stores where people in need could enroll more expediently for assistance. In general, the tone of industry and policy discourse about welfare applications of biometrics was consistent with Shoshana Magnet's argument that the rise of biometric identification of welfare recipients in the United States and Canada in the 1990s was a way of criminalizing the poor, especially poor women, and a development tied directly to the dismantling of New Deal welfare programs.[118]

The "war on drugs" and "welfare reform" programs had a close cousin: the war on undocumented immigrants. During the 1990s, neoliberal policies like NAFTA designed to break down economic barriers were accompanied by the buildup of policing along the U.S.-Mexico border. One of the strategies that state agencies employed to project an image of progress toward control of the border, and to manage "the twin policy objectives of *facilitating* cross-border economic exchange [while] *enforcing* border controls," was to expand and technologically improve on bureaucratic identification systems and their application at the border.[119] Part of this effort involved plans to integrate biometrics into the travel document issuance and border inspection processes.[120] During the 1990s, the U.S. Immigration and Naturalization Service (INS) began to introduce biometric systems for identifying legal visitors to the United States. In 1996, the INS commenced the *voluntary* INSPASS biometric system for frequent business travelers, and federal legislation made *mandatory* the use of biometrics with nonimmigrant temporary visas issued to Mexican citizens. INSPASS, short for INS Passenger Accelerated Service System, was instituted as a voluntary service available to businesspeople traveling to the United States three or more times per year, with citizenship from the United States, Canada, Bermuda, or one of the countries participating in the U.S. Visa Waiver program (mostly countries in western Europe). The mandatory temporary visas issued to eligible Mexican citizens allowed cardholders to make temporary trips across the border. These Border Crossing Cards (BCCs), also referred to as "laser visas," were issued to millions of Mexican citizens and contained individuals' biographical data, a facial image, two digital fingerprints, and a control number. In addition to storing the information and images on the card, the State Department created a database of facial images and other identifying information of the cardholders (later used as a sample database for government testing of commercial facial recognition systems).

The differences between the voluntary INSPASS and the mandatory biometric "laser visas" underscore the differential application of biometric technologies—not all uses, or *users*, are the same. Taken together, they present a more complete picture of the ways in which automated forms of identification over networks serve the neoliberal aims of mass individuation and social differentiation, a means of tying individuals into circuits of inclusion and exclusion, determining their status as legitimate citizens or more problematic identities deemed to fall outside the realm of democratic legal rights.[121] Biometric systems have been designed to fulfill the demands of institutional users to create greater ease of access for some, and more difficulty of access for others, depending on one's location in what James Boyle has called "the pyramid of entitlement claims" of the information society.[122] And while no overarching biometric system tracks individuals across institutional settings, these technologies are helping to approximate what Haggerty and Ericson call a "surveillant assemblage," a rhizomic system that "operates by abstracting human bodies from their territorial settings and separating them into a series of discrete flows," which are then "reassembled into distinct 'data doubles' [to] be scrutinized and targeted for intervention."[123] The "surveillant assemblage" allows for pervasive and relatively invisible ways of "distributing the living in the domain of value and utility," investing institutional users with a technical means of defining and differentiating the population and sorting out individuals and groups according to calculated levels of privilege, access, and risk.[124]

### *Obstacles to Institutionalization*

While it is important to recognize the particular appeal and demand for new surveillance technologies, it is likewise crucial to remain cautious about making facile assumptions about a pervasive, interconnected surveillance infrastructure capable of monitoring the entire society. Ubiquitous biometric identification has never been a foregone conclusion. References to the ease with which facial recognition technology could service the needs of businesses and public-sector users belied formidable challenges to implementing the technology and integrating it into administrative practices. The effort to build a new machinery of automated identification and mass individuation required not just technological ingenuity but also massive investments of political capital, economic resources, and human labor. Organizations considering facial recognition technology to meet their needs for securitizing identification systems had to consider the high cost and complexity of system integration. They also had to consider other problems, like building and maintaining databases, policies and procedures for information sharing, training workers and other end users how to operate the new technologies, the effects of error rates (false positives and false negatives) on system operation, and the risk of investing in a system that could conceivably become obsolete in a very short time.

One of the greatest issues impeding the adoption of biometrics, and one of the most pressing self-identified problems that the biometrics industry faced, was the lack of mature technical standards. Immature standards meant that potential users of biometrics could not easily replace one system with another, nor could they easily adopt multiple biometric systems to suit their needs. In the United States, assistance with negotiating technology standards is one of the major ways in which the federal government has supported biometric system development—through efforts of the National Institute for Standards and Technology (NIST), for example, as well as through organized partnerships with industry, such as the activities of the Biometrics Consortium, an industry-government association formed in 1992.[125] The politics and complexity of biometric standards setting provides clear evidence that by no means has the early effort to institutionalize facial recognition and other biometric technologies been a simple matter of plugging in shiny, new, seamlessly functioning devices.

The scale of effort required to build functioning facial recognition systems for institutional uses has presented perhaps the greatest challenge. The obstacles to full-scale implementation of facial recognition technology in driver's license administration, for example, were especially formidable given the size of DMV systems. DMV offices could control the conditions under which ID photos were taken, but the size of image databases would certainly put to the test the kind of one-to-many facial image searches that would be required to find stored facial images using the photos of new applicants and renewals. Using *images* as search criteria, rather than searching for words or numbers in the metadata attached to images, remained an extremely difficult technical problem, especially when the content was something as dynamic as the human face. And the larger the image database, the more difficult the challenge of content-based image retrieval (CBIR). As developers recognized, improving CBIR techniques would require experimentation with large-scale databases of faces in real-world settings, like those DMV offices were building.[126]

Immigration control systems presented similar challenges of scale. Although mandatory biometric registration of Mexican citizens applying for BCCs became a matter of policy, a digital regime of identification did not immediately take shape on the border between the United States and Mexico.[127] The ambitious, large-scale biometric identification systems never worked precisely according to plan. While millions of Mexican citizens obtained laser visas, most of the inspection stations did not install the equipment and software needed to read the digital identification documents, since Congress failed to appropriate the money for the biometric readers themselves.[128] Without the proper machinery in place, the laser visas were used for identity verification in the same way as conventional, non-machine-readable documents, with humans doing the perceptual labor of inspecting documents and visually comparing photos with faces. (The problems with fully implementing a laser visa system for Mexican visitors to the United States foreshadowed the problems with implementing the US-VISIT automated entry-exit system for *all* visitors, mandated after 9/11.) As the persistence of a documentary regime of verification at the border suggests, technological advancements in surveillance systems have not happened automatically, instead requiring considerable investment that is hard to sustain.

The deployment of ubiquitous biometric identification was far from fully achieved in the 1980s and 1990s, and in fact biometrics technologies remain experimental and sporadically applied in both public- and private-sector applications. The reasons for the "suppression of the radical potential" of biometric technologies have been complex and varied, including the need for technology standards, the enormous investment in resources and political will required, and the costs and major challenges of systems integration.[129] Another central issue has been the problem of "user acceptance": convincing "ID subjects" to overcome their "unfounded fears" of being subjected to biometric identification on a routine basis.[130] Labor control, according to David Harvey, has never been a simple matter of repressing workers, but instead has entailed "some mix of repression, habituation, co-optation and co-operation, all of which have to be organized not only within the workplace but through society at large."[131] As we will see in chapter 4, the institutionalization of biometric identification has required habituating individuals to the use of these technologies, and convincing them of the *need* to use them for their own "personal security."

## *The Eve of Destruction*

By the end of the 1990s, the biometrics industry appeared to be taking off, as a range of organizations attempted to integrate computerized facial recognition as well as optical fingerprinting, voice recognition, and related technologies into their identification systems. In January 2001, MIT's *Technology Review* named biometrics one of the top ten technologies that would change the world, and the biometrics industry revenues were estimated at $196 million for 2000, up nearly 100 percent from $100 million in 1999.[132] Facial recognition technology held only a small share of this market, but it held out unique potential for social actors interested in new ways of securitizing and standardizing identification systems. Despite the technical complexity of automated facial recognition and the scale of effort that would be required to integrate the nascent technology into surveillance and identification systems, the technology continued to capture attention and investment thanks to the central importance of the face and facial portraits in both the cultural and bureaucratic construction of identity.

Although it was impossible to foresee what was about to occur, the demand for more secure forms of identification was firmly entrenched in U.S. political culture on the eve of the September 11 terrorist attacks, virtually guaranteeing a response that would support more intensified biometric system development. Before turning to the attention that facial recognition technology received after 9/11, the next chapter considers a moment just preceding the attacks, in the summer of 2001, when Visionics Corporation partnered with the Tampa Police Department in an effort to bring the technology to an urban neighborhood in Florida. This project, the first of its kind in the United States, aimed to integrate automated facial recognition with a police video surveillance system being used to monitor the historic and entertainment district of Ybor City. It was conceived as a way of assisting the Tampa Police in performing the perceptual labor of surveillance, a means of making more productive use of an existing closed-circuit television system to target criminal identities in an area designated for urban renewal. But facial recognition technology, much like the urban renewal effort itself, ultimately promised much more than it delivered.

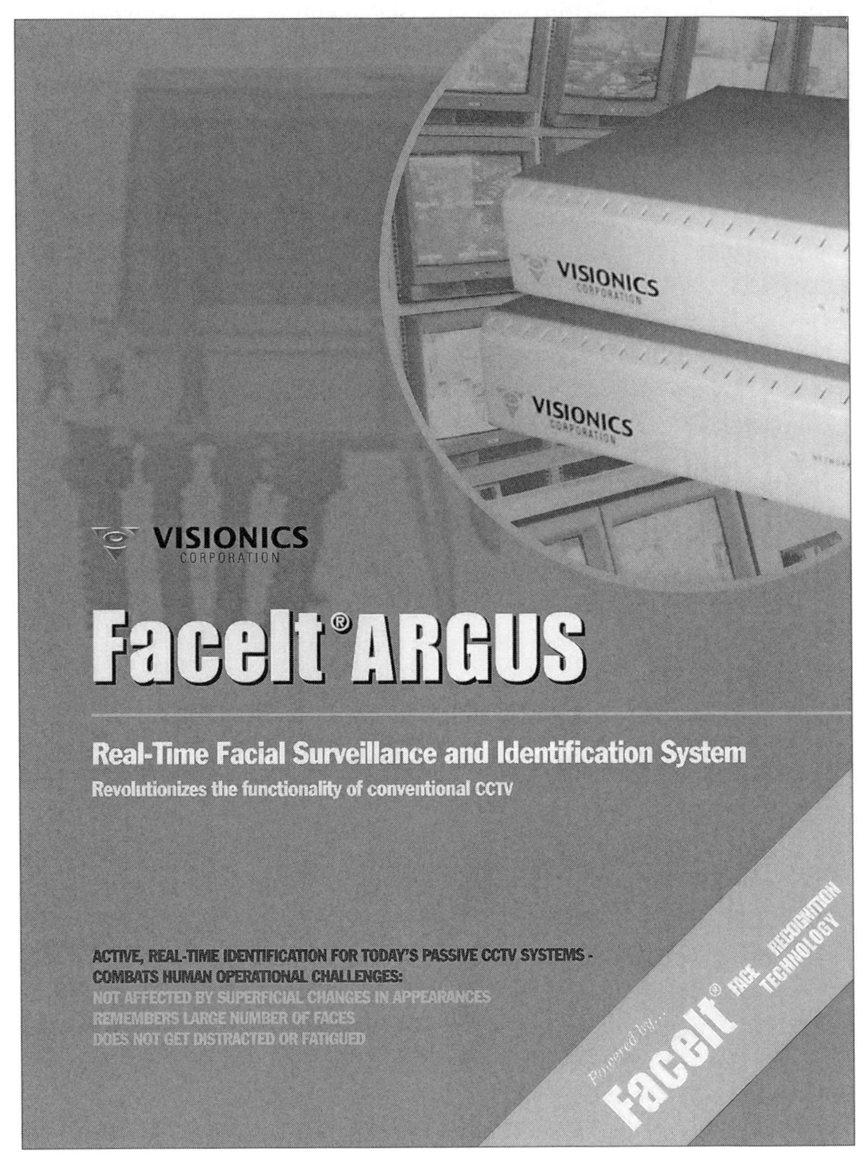


A brochure for Visionics' FaceIt® ARGUS facial recognition system for closed-circuit television systems.

# 2

# Police Power and the Smart CCTV Experiment

Real-Time Facial Surveillance and Identification System
Revolutionizes the functionality of conventional CCTV
Active, real-time identification for today's passive CCTV systems—
Not affected by superficial changes in appearances
Remembers large numbers of faces
Does not get distracted or fatigued.
—Visionics Corporation, brochure for FaceIt ARGUS, 2002

In her classic essay "Tales from the Cutting Room Floor" published in *Harper's*, Debra Seagal recounts five and a half months in 1992 that she spent working as a "story analyst" for the television show *American Detective*. Referred to in-house as a "logger," Seagal's job involved sitting in front of a "computer/VCR/print monitor/TV screen/headphone console," scrutinizing hours of raw video footage taken for the show, and creating a running log of all the visual and auditory elements that could be used to "create" a story.[1] Laboring at this monotonous job of video image processing, Seagal quickly descends into the abysmal world of the reality-based cop show, forced to watch the destitute and marginalized people who appear in the images before her—mostly prostitutes and people committing petty drug crimes—as they experience all manner of humiliations and injustices. She loses interest in the detectives whose sordid work she is obliged to witness, and begins to empathize with the "little people," "whose stories never make it past the highlight reel."[2] Although she did not identify much with the police, Seagal's job at *American Detective* bore some important resemblances to the work of certain police workers themselves, especially those unfortunate enough to have the post of monitoring video surveillance systems. In the 1990s, a growing number of people had to perform such labor, as the use of video surveillance systems by police and private security agencies became more widespread. This "visualization" of police work has had wide-ranging implications, beyond simply helping the fight against crime. Like Seagal's account of her brief career as a video analyst, the spread of police video surveillance has raised important questions about the labor of processing this explosion of visual information, about the amount and form of police power that video technology has enabled, and about struggles over the legitimacy of that power—all issues of central concern to this chapter.

While interest in the use of television technology for police surveillance is almost as old as television itself, it was more recently that the volume of such use began to pose major challenges for the police. As a result of their expanded use of closed-circuit television in the 1980s and 1990s, police and private security agencies began to face the problem of how to handle the explosion of visual information being generated by these systems. A single CCTV system, with twenty cameras running twenty-four hours a day, produced the equivalent of 480 hours of video footage, exponentially increasing the amount of work required to monitor these systems.[3] Surveillance and identification practices posed a problem of labor: not only was monitoring surveillance video mind-numbingly dull work, but the labor costs of managing thousands of hours of video was placing a heavy burden on both public law enforcement agencies and the growing private security sector. Recognizing a business opportunity, entrepreneurs attempting to commercialize facial recognition technology introduced the idea of "Smart CCTV"—the integration of automated facial recognition with video surveillance—as a potential solution to these problems of surveillance labor and video overload. Automated facial recognition, along with other computer vision technologies like license plate recognition and "anomaly detection," promised to provide a means of creating "algorithmic" forms of surveillance that could automatically manage the enormous amount of video imagery generated by CCTV systems without adding hundreds of human observers.[4]

But was facial recognition technology ready for such an application? What would have to happen to make Smart CCTV a reality, a functioning technology for monitoring city streets and other spaces? How would the technology be integrated into police practice, and what would that practice look like as a result?

In June 2001, Ybor City—a historic-entertainment district in Tampa, Florida, known as Tampa's "Latin Quarter"—became the first urban area in the United States to have its public streets fitted with a Smart CCTV system. Visionics Corporation began a project with the Tampa Police Department (TPD) to incorporate the company's automated facial recognition product, called FaceIt, into an existing thirty-six-camera CCTV system covering several blocks along two of the main avenues in Ybor City. Installed for free by Visionics, this new high-tech form of video surveillance promised to provide security for an area targeted for urban redevelopment, to help provide the security that was needed to transform central Ybor City into a more desirable tourist and consumer destination. The system was designed to automatically search digitized images of faces grabbed from video feeds against a watchlist database of wanted individuals, enabling the police to target those specific individuals for exclusion from the area (much like the unfortunate folks getting busted on reality-based cop shows). The "smart surveillance" system promised to benefit both Visionics Corporation and the Tampa Police, serving as an experimental test site for the company's technology and putting the TPD on the cutting edge of new police technology. The physical space of Ybor City, once known as the cigar manufacturing "capital of the world," would be transformed into a sort of "digital enclosure," a *virtualized* urban space, safe and secure for middle-class consumers.[5]

But the Ybor City smart surveillance experiment did not go as smoothly as its planners had hoped. The announcement of the system's installation triggered a heated debate, playing out on the streets of Ybor City, in the local and national press, and in the halls of the Tampa city government. Supporters claimed that facial recognition technology would help the police make Ybor City a safer place and thereby bring new life and business to the area, while opponents countered with accusations that it was too Orwellian and would ruin the unique and lively character of the neighborhood. Others suggested that the technology did not work and so was at best a waste of time and at worst a dangerous diversion of police resources. These competing claims about the system plagued efforts on the part of proponents to establish it as a necessary, desirable, and functional "security solution" for Ybor City. After a two-year free trial period, the Tampa Police abandoned the effort to integrate automated facial recognition with the Ybor City CCTV system in August 2003, citing its failure to identify a single wanted individual.

In this chapter, I argue that the failure of the Ybor City Smart CCTV experiment should not be viewed as solely a technical one, nor does it spell the end of the attempt to integrate facial recognition technology with video surveillance systems. Instead, the failure of the Ybor City experiment demonstrates that building a functioning Smart CCTV system involves resolving a complex combination of social and technological problems. The failure also reveals the questionable nature of claims that proponents made about both the maturity of the technology and its "technical neutrality." The failed Ybor City experiment suggests that, although proponents maintain that the technology is fully functional, in fact Smart CCTV is still very much a "technology in the making," and the ability to create computer programs that can reliably identify human faces from video feeds is not a foregone conclusion. Like all technological systems, Smart CCTV does not develop as an autonomous force moving forward of its own volition, but instead requires the concerted investment of a host of social actors, often with competing or incommensurate interests. Most important, the failure of the Ybor City experiment reveals the extent to which new surveillance technologies represent sites of struggle over the extent and limits of police power and over appropriate forms of governance in late capitalist, postindustrial urban spaces.

## *The Problem with CCTV*

Closed-circuit television is a transmission system for television that differs from the broadcast form typically associated with the popular medium: "Live or prerecorded signals are sent over a closed loop to a finite and predetermined group of receivers, either via coaxial cable or as scrambled radio waves that are unscrambled at the point of reception."[6] CCTV systems take many different forms and are used to perform a range of functions, from electron microscopy and medical imaging to pay-per-view sports broadcasts and live, on-site video displays for special events.[7] Although commonly viewed as a more recent phenomenon, police use of CCTV dates back at least to the 1960s in Britain.[8] But it was the 1980s and 1990s that saw an exponential increase in the use of closed-circuit television by police and private security firms in both the United States and Europe for monitoring urban spaces, gated communities, workplaces, and capital-intensive spaces such as banks, retail outlets, and casinos. Britain has far outpaced other countries in the extent of police CCTV deployments, but other countries have also experienced significant growth, especially in private-sector security applications.[9] In the United States, police in at least twenty-five cities had installed CCTV systems to monitor public areas by 2001, the year the Ybor City project began, and many more police departments were considering doing so in efforts "to give troubled down-town business districts a new lease on life, help public housing communities reduce destructive criminal elements, increase safety in public parks, monitory traffic congestion and catch red light violators."[10]

If one takes as given the role of the police as arbiters of law and order and believes that they should have wide latitude in performing that role, there seems little need to question the reasons for police adoption of new surveillance technologies, beyond concerns about their cost and effectiveness. Similarly, if one accepts assumptions about crime and criminality as being causes of social disorder rather than effects—the prevailing orientation that the police themselves take to defining the problem of crime—then the solutions rather obviously center on more police power, including more police surveillance. In his study of police power and cultural narrative in the twentieth-century United States, Christopher Wilson identifies a "paradox of modern American cultural life": "that much of our popular understanding of criminality and social disorder, particularly street disorder, comes from a knowledge economy that has the police—putatively agents of order—at its center."[11] If the study of the moral panic surrounding mugging produced by the Centre for Cultural Studies at Birmingham in the 1970s is any indication, the same holds true in Britain.[12] The prevailing police views about crime and disorder that have emerged in the United States and the United Kingdom since the 1970s are not especially sympathetic to arguments that challenge the authority of the police or offer broader social and political-economic explanations about the sources of crime and criminality. In the words of William Bratton, former police chief of the LAPD and former NYPD police commissioner under Rudolph Giuliani, "It is a great disservice to the poor to say that they lose jobs and so become criminals. . . . The penicillin for dealing with crime is cops. I thought I had already proved this. Criminologists who say it is economics or the weather or some other thing are crazy."[13] Bratton's comments express the predominant contemporary view of the police toward the problem of crime, a view (not entirely new) that dismisses social analyses of the "root causes" of crime as detached from the brutal reality of the streets.

The so-called realist view that stepped-up policing and surveillance is the solution to the "crime problem" not only shapes present police practice but also carries over into both public understandings of crime and policy orientations aimed at dealing with it, and the prevalence of this view makes it increasingly difficult to contest, in any meaningful way, police adoption of new "crime prevention" technologies. But not everyone agrees that increased police power is the answer to the crime problem. In fact, police power itself has long been a political issue in modern democratic societies, and not only among radical social critics. For example, the civil rights activist James Baldwin's charge that the police were the "occupying armies" of the inner city was taken up as a topic of urgent consideration by liberal reformers in the United States in the 1960s.[14] The debate about the legitimacy of police power and its appropriate limits is ongoing, if increasingly muted, and it represents one of the main reasons why the spread of CCTV systems has generated some, albeit minor, controversy. It also represents one of the main reasons why there has been opposition to the effort to technologically improve on CCTV, although the reason is not always explicitly stated in these terms.

Part of the controversy is sparked by the research of sociologists, legal scholars, and other critical observers, who have raised questions about the causes of CCTV proliferation and its social and political implications. According to this body of research, the seemingly self-evident reasons police give for adopting CCTV elide more complicated relationships between the spread of video surveillance, the social role of the police in modern societies, the material and symbolic functions of police power, and the social construction of crime and disorder. Attention to these more complex relationships provides a more adequate understanding of the reasons why police have adopted CCTV, and why they continue to adopt new technologies that promise to make CCTV a "smarter" and more effective surveillance technology.

A number of scholars maintain that the spread of CCTV surveillance is tied to a marked shift in approaches to crime control and criminal justice since the 1970s, specifically a movement away from penal welfare and rehabilitation, and toward more actuarial and punitive approaches. As David Garland has argued, crime prevention strategies now take high crime rates as a normal part of life, leading criminal justice systems to experiment with new ways of *managing* crime rather than assuming that crime can be reduced or even eliminated by addressing the social conditions that produce it.[15] High crime rates and the persistence of the crime problem in the face of what appear to be failed law enforcement and correctional strategies have created "new problems of legitimacy and new problems of overload" for criminal justice systems.[16] In turn, this normalization of high crime rates, and the exacerbating problems of police legitimacy and work overload, have led to the adoption of strategies of crime control that seek to off-load responsibilities for crime prevention onto individuals and non-state actors, Garland argues, making the avoidance of crime part of the responsibilities of each citizen and organization, part of the built environment, and part of everyday life.

As part of the move to make crime prevention commonplace in everyday life, strategic shifts in crime-control strategies have also included explicit efforts directed at measuring and managing levels of public fear and insecurity. In the 1980s police officials and policy makers in both the United States and Britain realized that public fear of crime was to some extent detached from actual crime rates, and so they began to take measures aimed at changing public perceptions, regardless of their impact on crime itself. The reduction of the fear of crime among certain publics became a "distinct, self-standing policy goal."[17] One result of this new orientation to crime control is that closed-circuit television systems now hover over urban centers and shopping malls as a matter of course, extending the gaze of police or private security throughout those spaces, with the visible presence of cameras often standing in for the authorities themselves. CCTV systems are used to target not only criminals and suspects, but also public perceptions about crime. In other words, at least some of the work that surveillance systems do is symbolic, tied to the symbolic authority of the police. The pursuit of both CCTV systems and new technologies for Smart CCTV must be understood as part of the more symbolic aim of creating the perception of stepped-up policing—the attempt to reduce fear of crime among preferred groups by investing police with an image of high-tech surveillance capacity.

The reinvention of the "community policing" paradigm in the 1990s likewise had improvements in citizen perceptions about policing and crime as one of its primary aims. As they took root in cities across the United States, the professed community policing initiatives "spawned larger, more ambitious experiments" involving the redesign of urban zones to enable more efficient police oversight, though actual police foot patrol played only a minor role.[18] Ironically, surveillance cameras themselves, along with the circulation of surveillance footage of crimes caught on video, contributed to a diffuse fear of crime that put pressure on police forces and other authorities to move forward with new technologies designed to make city streets and suburban malls seem safer.[19] "CCTV is now a standard technological response to the current media-fueled panic about crime and public order, and the collapsing sense of public safety and collective solidarity on our streets," writes Stephen Graham.[20]

But it would be a mistake to characterize CCTV systems as performing a strictly symbolic function. Surveillance cameras are not just for show—police in fact *use* CCTV systems—but the ways they use them rarely follow in lockstep with the intentions of policy or system design.[21] In a major study of thirty CCTV control rooms in a northern English city, Michael McCahill examined the way that various actors involved in using the interconnected systems interacted with one another, and through those interactions limited the potential surveillance capacity of system integration. In other words, through various forms of noncompliance or partial adherence to prescribed uses, humans more or less got in the way of realizing the full potential of integrated CCTV technology: "The human mediation of technology places limits on panoptic systems, because those responsible for operating them do not take on the essential properties of instantaneity and speed which characterizes the 'new' technology."[22] Lynsey Dubbeld has likewise studied the limitations of CCTV functionality, focusing not on the human operators but on the ways that material design and technological infrastructure mediate and place limits on the surveillance capacity of CCTV systems. In her study of a CCTV system in use in railway stations in the Netherlands, "targeted surveillance was made problematic as a result of the particular design of the control room . . . as well as by the capriciousness of technical artifacts central to the operation of the CCTV network."[23]

As a result of these combined human and technological limitations, CCTV systems have predictably fallen short of expectations, both in terms of their capacity to prevent crime in designated areas and in terms of their symbolic potential to reduce "fear of crime" and confidence in the police among desired members of the community. Police interest in new automated techniques that promise to help them make more effective use of CCTV systems stems in large part from these failed expectations. Rather than abandoning unsuccessful CCTV systems, social actors involved in their deployment and management have pursued other avenues to address their shortcomings, including the integration and computerization of CCTV systems. Once surveillance video systems become part of the material form of police practice, inefficiencies and other organizational problems that they introduce into the everyday work of policing become problems in themselves. As Ericson and Haggerty have noted, the ever-increasing workload of the police—especially the "paper burden" that accompanies their role as "knowledge workers"—leads police agencies to "search constantly for improved computer-based solutions" that promise to fulfill the practical needs of police work while also serving as a source of organizational legitimacy.[24] It would be consistent with this organizational logic, then, to find police turning to "improved computer-based solutions" in their attempt to deal with the growing *video* burden.[25]

## *Smart CCTV and Interpassive Policing*

Law enforcement agencies have long been involved in the rationalization of facial identification, and they were among the primary early markets envisioned for automated facial recognition technologies. As a result of their critical need to manage volumes of photographic images, law enforcement agencies have been important sites of innovation in photographic archival systems, as Allan Sekula has shown.[26] The recent transition of police photographic practices from analog to digital, and from filing cabinet to database, is often overlooked in discussions of new developments in photographic media. As we saw in the previous chapter, police municipalities and prisons began adopting computerized mug shot booking systems on a wider scale in the 1990s, with companies like Digital Biometrics and Identix adding electronic mug shot capture stations to their live-scan electronic fingerprint systems. The federal government was also involved in the process of computerizing mug shot booking and criminal identification. In 1995 the FBI sponsored a Mug Shot and Facial Image Standards Conference, for example, in order to develop criteria for the kinds of images that the new law enforcement mug shot systems should be designed to capture, standards that would also "take account of emerging technologies like software that determines if two facial images belong to the same person."[27] Recognizing law enforcement agencies as a major market, facial recognition companies partnered with other IT companies like Sun Microsystems, ImageWare Inc., and Printrak International to integrate their facial recognition technologies with image processing systems already being marketed to law enforcement agencies. Soon thereafter, vendors of facial recognition and finger imaging began to merge to form companies that combined these product offerings. (For example, Digital Biometrics merged with Visionics in 2001 and then Visionics merged with Identix in 2002.[28])

But the automated recognition of faces from *surveillance video*, with its notoriously poor quality images, still presented considerable challenges. The FERET government tests of facial recognition algorithms conducted in 1996 showed that dynamic image matching and one-to-many searching of variable quality images resulted in much lower accuracy rates for facial identification. Follow-up testing in 2000, sponsored by the same agencies but renamed the Facial Recognition Vendor Test (FRVT 2000), confirmed that the new commercially available systems from companies like Miros, Visionics, and Viisage still had considerable limitations with respect to matching facial images taken from video.[29] The performance of facial recognition systems was affected by things like facial pose variations, the amount of time that elapsed between the original facial image and the probe image, the distance between the person and the camera, variations in facial expressions, and especially changes in lighting.[30] Other studies likewise consistently found that the quality of images taken from surveillance video was too variable to support reliable automated facial identification. In short, computers were well on their way to accurately matching faces in standardized photos taken in controlled settings, but still not doing very well at identifying faces from video taken in real-world surveillance scenarios.

Nevertheless, companies like Visionics were eager to move forward with the integration of facial recognition technology with video surveillance, recognizing a potential source of revenue in the critical need for the police to better manage and make more effective use of CCTV systems. Police needed new ways of managing exploding volumes of surveillance video and making more effective use of CCTV systems. They also needed to maintain legitimacy and symbolic crime-fighting capital by appearing on the cutting edge of new crime-control technologies. In the late 1990s, Visionics Corporation began developing and marketing their FaceIt facial recognition product as a solution to the twin problems of video overload and police legitimacy, claiming that the technology was an improvement over both existing, "passive" CCTV technology, and over the human monitors of CCTV systems. With each new press release, Visionics declared their FaceIt product better able to handle larger databases and greater numbers of facial images, faster and with more accuracy than previous versions. In October 1997, Visionics announced the release of "FaceIt Multiface™," ostensibly "the world's first face recognition system capable of capturing and identifying multiple faces in the same field of view, and tracking these faces continuously."[31] According to the announcement, FaceIt Mutliface could identify several people walking in a group without requiring them to stop or form a single-file line. (The system could capture faces from only a single video stream at a time, still falling short of the kind of recognition capacity needed to process video streaming from twenty-five, fifty, even two hundred cameras at once.) Less than six months after announcing the release of FaceIt Multiface, Visionics released another new version of FaceIt, called "FaceIt DB." CEO Joseph Atick claimed that FaceIt DB had the "ability to check every face that appears in a camera's field of view in real time," taking automated surveillance and identification "to its highest level, breaking new ground in the use of technology to fight crime."[32] Visionics later announced that they had refined their technology so that it could handle "changes in lighting, pose, expression and lighting conditions"—all factors that continued to pose challenges for automated facial recognition.[33] In 2001, Visionics released their "FaceIt ARGUS" system, declaring it "the first commercially available facial recognition product that can handle *an unlimited number of camera inputs* and provide real-time identification."[34] A brochure for FaceIt ARGUS (cited at the outset of this chapter) claimed that it "revolutionizes the functionality of conventional CCTV," provides "active, real-time identification for today's passive CCTV systems," and "combats human operational challenges: not affected by superficial changes in appearance, remembers large numbers of faces, [and] does not get distracted or fatigued."

The claims Visionics made about the video surveillance applications of their product were revealing about what proponents wanted to achieve with

Smart CCTV, if not what could be realistically accomplished. Visionics posited FaceIt as an improvement over both "passive" CCTV systems and the inefficient, fallible human operators of those systems. They pitched their technology as a laborsaving device, promising to save CCTV operators hours of time observing surveillance video and relieving them of the responsibility for identifying criminals and suspects that might appear on the screens. At the same time, there seemed to be an implicit acknowledgment that the kind of labor it would save the police and other users of CCTV systems was never possible in the first place. Facial recognition technology promised to quite literally do the watching *for* the CCTV operators, relieving them of the need to pay attention to the screen.

The possibility of delegating responsibility to the Smart CCTV system for the perceptual labor of watching video and recognizing faces suggested a form of passive media activity, one that Slavoj Žižek has referred to as "interpassivity."[35] According to Žižek, "interpassivity" is the uncanny supplement to the more fashionable and celebrated notion of "interactivity" associated with new media technologies. Whereas interactivity implies a spectator-user actively engaged with electronic media and taking part in the production of content and meaning, interpassive arrangements allow the medium itself to do the work of reception for the user. Žižek uses the example of the VCR aficionado (the compulsive DVR user is even more apropos) who records hundreds of hours of movies and television shows, knowing that there will never be time to watch all of it. Gijs Van Oenen has applied Žižek's concept of interpassivity to analyze the domains of labor and politics, drawing as well on Robert Pfaller's related notion of the *dromenon*, or the "machine that runs itself."[36] What Van Oenen calls the "interpassivization" of labor involves the automation of both manual and mental activity. Today, "hands-on" work means manipulating a computer interface, Van Oenen argues, and the prevailing tendencies of contemporary work arrangements make workers more alienated than ever from the products of their labor. The interpassivization of labor is deeply embedded in post-Fordist forms of labor organization, including outsourcing, more "flexible" workforces, and loose, "network" forms of business restructuring. These political-economic developments have had a profound effect not only on work arrangements but also on worker subjectivity, as workers are forced, paradoxically, to become both more flexible and more passive—to be prepared for constant technical retraining, relocation, and experimentation, while allowing machines to perform not only the manual but also much of the mental labor for them.

The promoted capacity of FaceIt to make "passive" CCTV systems more "active" and relieve human operators from their perceptual labor embodied this logic of interpassivity, suggesting that the solution to the problems of CCTV monitoring could be found in the interpassivization of police surveillance labor. The hands-on work of monitoring surveillance video—itself already a mediated form of police supervision—would involve merely responding to computer programs that would do the actual work of identifying dangerous threats to the community. If Smart CCTV worked, the human labor of monitoring would require less in the way of specialized police knowledge of criminal identities. This removal of human perceptual capacity from the process of identification was posited as a special benefit, a technically neutral form of identification that would ostensibly counter the prejudicial tendencies of police officers. Not only was the technology tireless, efficient, and memory-intensive, it also promised to function in a culturally neutral way, blind to racial or ethnic differences of faces. In the words of Visionics CEO Joseph Atick, his company's product "delivers security in a non-discriminatory fashion. FaceIt technology performs matches on the face based on analytical measurements that are independent of race, ethnic origin or religion. It is free of the human prejudices of profiling."[37] "Interpassive surveillance"—allowing facial recognition technology to perform the mental labor of watching—would ostensibly bring a measure of objectivity to police surveillance practices.

Of course, this is what was *promised* of Smart CCTV, not what it delivered. The claims that Visionics made about FaceIt—which were more than a little overstated—created expectations that would inevitably go unfulfilled. And whether in fact the technology would fit into the culture and practices of law enforcement remained to be seen. As William J. Bratton stated emphatically, the "penicillin" needed to deal with crime was more *cops*, not more technology. If the Ybor City Smart CCTV experiment is any indication, the cops themselves—at least the ones responsible for trying to make the facial recognition system work despite its technical limitations—were not necessarily ready to rush into the brave new world of interpassive policing.

## *The Ybor City Smart CCTV Experiment*

The first urban center to integrate facial recognition technology with a police video surveillance system was not the Ybor City neighborhood in Tampa, Florida, but the London borough of Newham. In 1998, Visionics Corporation partnered with a British-based company called Software and Systems

International to upgrade Newham's extensive CCTV system of 140 fixed cameras and 11 mobile units. "We have listened to the people of the borough and we have acted," said Newham councilor Ian Corbett. "Crime prevention is the number one issue for 60 percent of the people in Newham."[38] From the beginning, the politicians and police officials responsible for the Newham Smart CCTV project were concerned with creating the appearance of high-tech police surveillance as much as actually providing a means of apprehending criminal suspects. In explaining the need for the system, Robert Lack, Newham's security chief, pointed to problems of unemployment and increasing crime levels following the closure of the docks.[39] "The need was to reduce the *public fear of becoming a victim of crime* and increase the *criminals' perception* of the chance they would be detected," said Lack. "Back in 1996 public fear of crime across the Borough was above 75%; it now stands at 67% and is dropping."[40] As the collection of these statistical measurements of fear suggests, the actual effectiveness of the new Smart CCTV system would not be gauged strictly in terms of its ability to identify and intercept suspects, but also in terms of its effects on public perceptions of both crime and police surveillance. Whether the facial recognition system actually worked in practice would be less important than whether people—innocents and criminals alike—actually believed that it worked. As planned, the system would initially be used to identify muggers and shoplifters, including "members of a shoplifting ring nicknamed the 'Kenya Boys' by the local police," and eventually expanded to include "known or suspected pedophiles."[41] According to a spokesperson from the firm that designed the system, the technology had distinct advantages over human operators: its eyes never got tired after staring at screens for hours, and "it never goes to the loo, either."[42]

It took almost three more years for the police in a U.S. city to begin a similar project aimed at compensating for the limitations of human CCTV operators, and when the Tampa Police made the decision to try out the technology it was on a considerably smaller surveillance apparatus. The Ybor City CCTV system included only thirty-six cameras, and the first installment of the FaceIt system could run on video from only one camera at a time, requiring operators to choose which video feed to process. The story of how Ybor City became the first urban space in the United States to be fitted with Smart CCTV is, on one level, a story of business relationships and individual entrepreneurial initiative. The actual proposal to use FaceIt in Ybor City was initiated by Detective Bill Todd of the Tampa Police Department in early 2001, following a test of the Viisage facial recognition system during Super Bowl XXXV at Tampa's Raymond James Stadium. Detective Todd had worked as part of the plain clothes detective unit at the game, and he had a previous working relationship with David Watkins, then managing director of a company called Graphco Technologies, a "value-added reseller" of facial recognition systems, and the party responsible for the hands-on work of installation.[43] Watkins was an entrepreneur and technical communications specialist with many years experience developing law enforcement information systems. He had an interest and considerable expertise in designing networks to facilitate the sharing of investigative information among police departments, and the Ybor City facial recognition experiment would serve as a staging ground for the benefits of such networks, especially if wanted criminals from other districts were identified and apprehended on visits to the neighborhood. On the initiative of Watkins and Detective Todd, the Tampa City Council approved a one-year contract with Visionics to begin in June 2001. (The contract was later extended to two years.) The decision to choose Visionics technology, according to Detective Todd, was based on the most recent government testing, the FRVT 2000, where the Visionics FaceIt system outperformed other commercially available facial recognition systems.

Why Ybor City was chosen over other neighborhoods in Tampa for the Smart CCTV experiment stemmed from a number of converging factors. In many ways, Ybor City represented an ideal test site for such an experiment—David Watkins called it his "living laboratory."[44] Most important, the police were already operating a closed-circuit television system in the area, installed as part of stepped-up security initiatives that accompanied redevelopment projects in Ybor City in the 1990s. The neighborhood also had a high crime rate relative to other neighborhoods in Tampa and a bustling weekend party scene that gave it a reputation for being a risky place to visit. According to Detective Todd, police use of the new high-tech surveillance technology would "send a message" that they were "committed to enhancing the quality of life in our neighborhoods" and "making Ybor City a desired destination point for our citizens."[45] Much like the Newham Smart CCTV experiment, in Ybor City the police had in mind not only identifying criminal suspects, but also conveying an impression about the active role of the police in adopting new crime-fighting tools in order to make Ybor City a safe and secure place, a "desired destination point" for the mobile consumer.

The story of how Ybor City acquired its party reputation offers some insights into how and why it became the first public test site for Smart CCTV in the United States, and why the fate of the neighborhood seemed to become indelibly tied police surveillance (and not necessarily to its advantage). The area known as Ybor City, located on the eastern side of Tampa, was founded in 1886 at the beginning of a major wave of immigration from southern and eastern Europe to the United Sates. Named after the Spanish cigar magnate Vicente Martínez Ybor, Ybor City became home to a thriving cigar industry. Jobs in the cigar factories brought immigrant groups of Cubans, Spaniards, Sicilians, and Italians to Ybor City to work and live in Ybor's planned community.[46] The cigar industry declined by the mid-1930s, but Ybor City remained a stronghold of the same groups that had been drawn there by the cigar trade. It was the social, political, and economic changes following World War II, including activist federal government policies supporting urban renewal, that radically altered and disrupted the unique immigrant community of Ybor City.[47] Tampa's first urban renewal agency commenced operations in 1962 with a charter to rehabilitate and redevelop "slum areas." Bulldozers began tearing down Ybor City in 1965, preparing to create "a tourist attraction second to none in the U.S.," according to the urban renewal office.[48] Soon thereafter, newly constructed Interstate 4 split off Ybor City from its northern section. At least 1,200 families were displaced, leading to a downward spiral of neglect and urban decay. Civil unrest followed the police shooting of a young black man in Tampa in June 1967, and although Ybor City was not the site of the protests, media coverage had a damning impact on area, drying up support for urban renewal programs. Similar programs in another Tampa neighborhood known as "the Scrub" displaced a large number of low-income African American families, many of whom moved into vacant housing in Ybor City. As financial institutions redlined Ybor City, the razed land remained vacant, and the blight of the area deepened.[49]

By the late 1980s, however, a new, hip, bohemian culture was emerging in Ybor City, with a critical mass of artists attracted there by the cheap rents and unique, old-urban character of the neighborhood.[50] As one local reporter observed, after two decades of failed renewal efforts Ybor City seemed "to be getting a new lease on life," as the area began to draw tourists and local Tampa residents attracted to the artist studios, street vendors, and live entertainment.[51] The distinctive community taking shape in Ybor City began to attract the attention of developers. In an editorial titled "Ybor City's Past Can Enhance Tampa's Future," a local corporate lawyer advocated for renewed investment in Ybor City in order to help Tampa "become a world-class convention city" that could "compete with Orlando, Miami, and other Southeastern cities for convention business."[52] The Tampa city government stepped in, renewing its efforts to remake Ybor City, designating it a "Community Redevelopment Area" in 1988; two years later, Ybor City was named a National Historic Landmark District, one of only four in Florida. The Ybor City Development Corporation (YCDC) was established with the stated aim of serving as the "primary vehicle for planning, promoting development, and assisting with the redevelopment of Ybor City through public and private sector offices."[53] The YCDC devised a plan "to guide future development so as to eliminate existing conditions of blight" and encourage "the rehabilitation of the district through the stimulation of the private sector investment and business activity."[54] A series of development projects were undertaken in the area, the most significant of which was a fifty-million-dollar project called "Centro Ybor," a retail-entertainment complex in the heart of Ybor City. The residential areas within the Ybor City area included a combination of dilapidated housing, restored homes, and new condo developments, and the neighborhood remained bordered to the west by a government-subsidized housing project. Writing in 1995, the ethnographer Timothy Simpson commented on the cultural climate that had emerged in Ybor City, after the cycle of failed urban redevelopment programs and renewed efforts at preserving the unique heritage of the neighborhood:

> Ybor City is currently caught in the tension between being a district marked by "historical preservation" and being self-consciously in a "state of transition." . . . Nouveau art boutiques and trendy restaurants compete for attention with the boarded-up buildings and crumbling facades that surround them. . . . The cigar industry has all but disappeared from Ybor City, only to be replaced by the tourist industry, and unemployed residents and homeless people walk the streets alongside yuppie lawyers, tourists, and young Bohemians. The air is charged, though, with the possibility of community, of radical change, a moment that marks Ybor City's past yet seems to hover there in the present space of the city street.[55]

This charged moment of possibility might have blossomed if the needs of the neighborhood's local inhabitants were not subordinated to the imperative of making Ybor City a competitive convention, tourist, and consumer "destination point." In addition to the redevelopment projects that were reshaping the urban space of Ybor City, the City of Tampa introduced a set of incentives to entice businesses to locate there.[56] Rules requiring bars to be one thousand feet apart were suspended, and other standards governing storm-water drainage, parking provision, and transportation impact fees were also waived. The business sector that was most attracted by these incentives was the bar and nightclub industry, according to Eric Snider, a local reporter. Bar owners appealed in droves to the Tampa City Council for "wet zonings," permits that allow alcohol sales, and "the council complied, handing out wet zonings like Jolly Ranchers on Halloween."[57] While the alcohol permits spawned renovations to buildings that might have otherwise remained vacant, the result was the overproduction of bars in Ybor City. "Around the turn of the millennium," according to Snider, "Ybor City no longer deserved to be called an entertainment district. A drinking mall, maybe." At the same time, the Centro Ybor complex had managed to attract chain stores like American Eagle Outfitters and Pacific Sunwear, and needed to attract additional businesses in order to become a profitable retail center. In an effort to "clean up" the area and make it more hospitable to corporate retail establishments, the city ousted the small vendors and street performers who populated the streets in the 1990s, a move that destroyed the bohemian, artistic vibe and "sucked some of the freaky character out of the strip."[58]

What occurred in Ybor City during the last decade of the twentieth century resembled similar neoliberal socioeconomic transformations that occurred in cities across the United States: the redesign of urban public spaces according to corporate-defined redevelopment priorities, leading to the over-investment in retail-entertainment districts. As a result of economic crises in the 1970s, cities were forced to adopt a heightened competitive posture, vying for position as centers of consumption, among other dimensions of strategic competitive advantage.[59] This competitive stance was particularly intense in Florida, a state whose economy depends heavily on tourism and convention business. Tampa was in constant competition for tourist and visitor dollars with other Florida cities, including Orlando, just eighty miles east. (Within cities, one shopping center after another competed for the business of visitors and local suburbanites. Ybor City competed on a local scale with other retail and entertainment complexes within Tampa and in neighboring St. Petersburg.) In the course of Tampa's effort to gain an edge in the context of heightened interurban competition, efforts were made to remake Ybor City into a "variation on a theme park," Michael Sorkin's phrase for privatized spaces of consumption designed to capitalize on a nostalgic, stylized, and commodified version of the past.[60] While urban planners seemed especially concerned with preserving Ybor City's distinctive "Latin" heritage, others saw the revitalization projects as decidedly business- and profit-oriented, with the Latin theme taking a marketable, Anglo-friendly form.[61] What consistently accompanied these commercially oriented urban "revitalization" and "renewal" programs, as Mike Davis observed of Los Angeles, was an "obsession with physical security systems," and "an unprecedented tendency to merge urban design, architecture and the police apparatus into a single, comprehensive security effort."[62]

Given this concerted, if flawed effort to remake Ybor City into a tourist-consumer mecca through a model of competitive, privatized urban redevelopment, it was not surprising to find the Tampa City Council and the Ybor City Development Corporation moving in 1997 to direct public funds for the installation of a closed-circuit television system in Ybor City, to be monitored by the Tampa Police. The area would have to be purged of its undesirable inhabitants and visitors if it was ever going to be a place where people with money would come to spend it. And when the neighborhood inevitably failed to generate sufficient consumer dollars to support the demands of corporate-defined redevelopment projects, the blame was consistently placed on the problem of crime and public perceptions of the area as too dangerous to visit. Adopting a strategy that was already widely deployed in retail and other private establishments and in public spaces all over Britain and in select U.S. cities, the Tampa Police installed surveillance cameras covering the busiest blocks along two of the main avenues in Ybor City.[63] Only one Tampa City Council member voted against installing the CCTV system, saying that he did not think there was "a compelling enough reason to 'whittle away' at the public's freedom of movement by recording what bars they frequent or which people they meet."[64] Tampa deputy police chief John Bushell disagreed: "This isn't a Big Brother kind of thing. . . . We just want to make it a place where people can come and feel comfortable walking around at night."[65]

Here once again, the police voiced as one of their primary concerns the need to reduce the public's *fear of crime*, to deter certain people from the area and assure desired visitors about the safety and security of "revitalized" urban space. Chief Bushell went so far as to tell the *Tampa Tribune* that most of the time, no one would be watching the surveillance video; the cameras were to be left in recording mode except during peak hours on weekend evenings, when an officer would man the monitors in real time.[66] In an editorial published a year before the CCTV system was installed, the *Tampa Tribune* itself came down firmly in support of the plan: "Any doubts about the need for surveillance in Ybor City should have been blown away by the gunfight on crowded Seventh Avenue early Sunday morning. . . . The cameras are no replacement for a strong police presence, but they should increase the officers' effectiveness."[67] As the *Tribune* editorial suggested, Ybor City was struggling with a rough reputation as a crime-ridden area, an image that was seen as especially detrimental to the transformation of the neighborhood into a "destination point" for mobile consumers from the suburbs, as well as tourists and visitors attending conventions in Tampa.

The decision to experiment with facial recognition technology in Ybor City was consistent with the logic governing the earlier adoption of the CCTV system, and similarly tied to concerted attempts to remake the area into a profitable version of the past, a space of consumption and leisure for a critical mass of people with buying power. But in addition to a preoccupation with reducing the fear of crime and making shoppers and tourists feel safe and secure, one of the main reasons the Tampa Police repeatedly gave for adopting the original CCTV system, as well as the upgraded Smart CCTV technology, was the need to monitor the crowds that gather in Ybor City on weekend evenings. One of the greatest "operational challenges" police faced, according to Detective Todd, was dealing with the crowds of revelers that spilled out into the streets on Friday and Saturday nights, creating a chaotic scene that pushed beyond the boundaries of policing capacity.[68] Every weekend evening the crowd would explode with partying enthusiasm, but it

A surveillance camera mounted to oversee the Centro Ybor entertainment complex in Ybor City. Courtesy *Tampa Tribune*. Reprinted with permission.

Surveillance signage on Seventh Avenue in Ybor City. Courtesy *Tampa Tribune*. Reprinted with permission.

always threatened to become a major problem, or many minor problems that overpowered the police officers assigned to patrol the area. Particularly in the discussions about the need for facial recognition technology, the crowd was described as riddled with "dangerous individuals"—thieves, drug dealers, and especially "sexual predators"—who eluded the police and preyed on the innocent. With crowds numbering as many as thirty thousand people, according to Detective Todd, "traditional police tools break down. . . . Patrol officers walking in that crowd have trouble seeing what's going on," so the police "began mounting cameras on light poles to give police officers a better view of the street."[69]

The problem of the crowd in the city has a long genealogy, accompanied by the corresponding development of technologies and spatial arrangements designed to govern the space of the city and make the occupants of the crowd more visible. For nineteenth-century social theorists like Le Bon and Sighele, the crowd embodied "a highly emotional, irrational, and intolerant 'psychic current' . . . the lowest form of 'common consciousness.'"[70] The specter of the crowd that haunted social thought inspired "the invention of technologies of spaces and gazes," explains Nikolas Rose, "the birth of calculated projects to use space to govern the conduct of individuals at liberty."[71] The crowd posed a constant threat to public order, and so necessitated new spatial arrangements and codes of conduct designed "to transform towns from dangerous and unhygienic aggregations of persons into well-ordered topographies for maintaining morality and public health."[72] Town planners envisioned the construction and maintenance of the healthy "liberal city" through the orderly arrangement of public spaces in ways that opened them up to visibility and made each individual the target of "a play of normative gazes," under close observation not only of the authorities but also of one another.[73] In his 1978 lectures on security, territory, and population, Foucault similarly addressed the moment when economic development made necessary the suppression of city walls, reducing the supervision and control over daily comings and goings of city occupants and thereby generating new insecurities from "the influx of the floating population of beggars, vagrants, delinquents, criminals, thieves, murderers, and so on" who came from outside the city.[74] The lack of physical barriers around the city necessitated new ways of making the space visible and maintaining control over the bodies mingling in that space.

The police attention to the problem of the crowd in Ybor City, and their interest in surveillance technologies designed to make the crowd more visible and controllable, suggests a new manifestation of these earlier efforts to construct the "liberal city." The Tampa Police expressed similar concerns about the crowd and the threat it posed to the orderly maintenance of urban space. The physical presence of the police was augmented by the presence of cameras and signage throughout the space of the neighborhood, but the arrangement of video surveillance was soon deemed less than optimally effective, since the "floating population" continued to penetrate the space of the crowd, crime rates continued at unabated levels, and the redeveloped space failed to generate profit. Facial recognition technology promised to perform some of the labor of monitoring, breaking through the anonymity of the crowd to target select individuals. On a symbolic level, the publicity surrounding the facial recognition system was meant to convey the message that individuals would no longer be anonymous in the crowd, but instead would be individuated and identifiable. As in Foucault's plague-stricken town, the compact mass of the crowd would be "abolished and replaced by a collection of separated individualities."[75] The accumulation of coded information about criminals and criminal suspects would be brought to bear on the entire space of Ybor City, after which appropriately behaved, law-abiding consumer-citizens would have nothing to fear from the "floating population" of threatening individuals. Nor would well-behaved citizens have to fear police disruption of their free movement and activity, since the facial recognition system would accurately and authoritatively pick out problem identities through precise, statistical processes of classification and individuation. In this way, the Smart CCTV system promised to transform Ybor City into a "digital enclosure," converting the physical space into scannable content that could be enfolded into the virtual space of computer networks and made subject to their monitoring capacity.[76] In the digital enclosure, individuals could be appropriately scanned, identified, and classified according to automatic determinations of whether or not they belonged, and excluded from the space if deemed a threat. Ease of access to and through central Ybor City would be facilitated for civilized, law-abiding consumer-citizens and restricted for derelict and suspect individuals.

Of course, the promise of Smart CCTV to automatically sort out the civilized from the uncivilized glossed over the human decisions that go into making determinations about who belonged and who did not. How would this automatic sorting scheme be accomplished? The ability of the facial recognition system to individualize the crowd and target problem identities would depend significantly on the construction of an electronic archive of suspect individuals, also known as a watchlist database. The contents of the watchlist database, the basis of the system's memory, would determine the range of faces it could conceivably identify. In order for the Smart CCTV system to identify an individual face, it would translate an image of a face into a smaller amount of data (called a "facial template" or "faceprint") that could then be matched against a database of facial images.[77] But precisely what criteria would govern the composition of the watchlist database and how would information about suspect identities be shared and managed? What identities would be included in the database and for how long? What specific information about these individuals would be archived? How would that information be organized, and how would the accuracy of the information evaluated? These were policy questions, as well as questions about the practical design of the identification system. In order to understand how an automated facial identification system would function, it is crucial to examine the social forces governing the form and content of image databases. As we will see in chapter 3, the construction of a watchlist database involves forms of *social classification*, where particular assumptions about dangerous identities, including long-standing race- and class-based assumptions, unavoidably get designed into automated surveillance and identification systems.

In order to invent the accuracy of automated facial identification and position it as a functioning technology, Visionics elided important questions about database construction, as well as questions about the maturity, feasibility, and desirability of the technology.[78] In a press release announcing the installation of Smart CCTV in Ybor City, CEO Joseph Atick claimed that "the growing acceptance of the use of FaceIt® technology in CCTV control rooms validates it as an important, powerful tool for law enforcement."[79] It is unclear where Atick found this "growing acceptance" or why such acceptance would validate the usefulness of the technology for law enforcement. Ironically, the press release did include the following disclaimer, as per policy of the SEC: "This news release contains forward-looking statements that are subject to certain risks and uncertainties that may cause actual results to differ materially from those projected on the basis of such forward-looking statements." Included as a message to the intended audience of potential investors, the disclaimer was meant to warn them not to take any of the promises made in the press release as absolute facts about the company's future growth potential. But while the company continued to experience growth (thanks primarily to its successful digital fingerprint systems and mergers with other biometrics companies), the message was especially apropos with respect to the claims Visionics made about the Ybor City facial recognition experiment, where the actual results would in fact "differ materially from those projected." The announcement was not entirely forthcoming with investors, however: it did *not* mention that Visionics installed the facial recognition system free of charge, instead implying to the investor community and to potential buyers that the Tampa Police Department was a paying customer.

The Visionics press announcement also seemed to anticipate the objections of opponents. The company was careful to explain that the system was designed "in accordance with industry-established privacy guidelines and with existing law governing the use of personal data by public agencies." Not only did FaceIt work perfectly well with video surveillance, according the official statement, but apparently all the necessary legal and policy issues had already been resolved. Visionics was a responsible company complying with the law and with industry self-regulation with respect to the privacy of personal data, so there would be no need to worry about the deployment of automated facial recognition for police surveillance of public streets. But as it turned out, not everyone was willing to accept the self-professed responsibility that Visionics claimed to assume, and the company was right to anticipate opposition to the Ybor City Smart CCTV experiment.

David Watkins, the technical specialist who installed the Visionics FaceIt system in Ybor City, demonstrating the system to the press in a Tampa Police CCTV control room. Courtesy *Tampa Tribune*. Reprinted with permission.

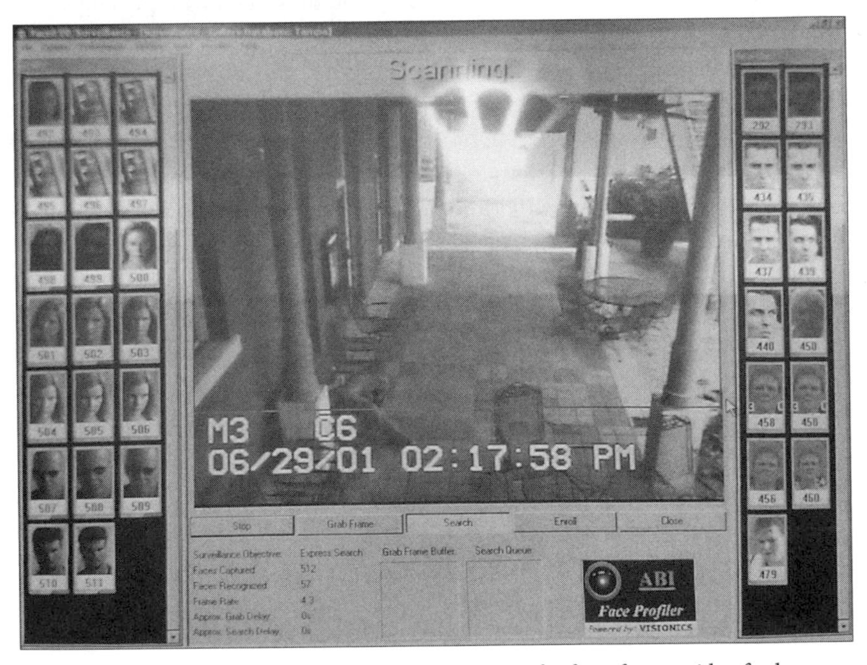


A screen image of the Visionics FaceIt system scanning for faces from a video feed. Courtesy *Tampa Tribune*. Reprinted with permission.

## *The Controversy over Smart CCTV in Ybor City*

No sooner had Visionics announced the installation of FaceIt than a heated "war of interpretations" broke out over police use of the new surveillance technology in the public streets of Tampa. The announcement of the Tampa Police Department's plans to use facial recognition technology in Ybor City attracted considerable attention from the local and national press, from the ACLU, from policy makers, and from other individuals and groups representing a range of interests. Rather than seeing this debate as circulating outside the project itself, it is important to recognize that the competing interpretations of the technology would themselves have a role in shaping the form that automated facial recognition would take in Ybor City, or whether it would take any form at all. "With a technological project," writes Bruno Latour, "*interpretations* of the project cannot be separated from the project itself, unless the project has become an object"—in other words, unless the project has taken shape as a functioning system.[80] Precisely how the war of interpretations over the technology played out would have a determining influence on whether in fact Smart CCTV became a reality in Ybor City.

While the local Tampa press began reporting some unease among people on the street in its early news coverage, it was the decision of U.S. House

Protestors demonstrating against the installation of facial recognition technology in Ybor City. Courtesy *Tampa Tribune*. Reprinted with permission.

Majority Leader Dick Armey (R-TX) to join the ACLU in public opposition to police use of facial recognition technology that put Ybor City experiment on the national stage. "I'm not sure there's been a case so perfectly Orwellian," declared Rep. Armey, adding that "placing police officers in a remote control booth to watch the every move of honest citizens isn't going to make us safer."[81] Randall Marshall, the legal director for the American Civil Liberties Union of Florida, argued that it amounted to placing everyone on the street in a "virtual lineup," and that more public deliberation was needed before police moved forward with adopting the technology: "This is yet another example of technology outpacing the protection of people's civil liberties. . . . It has a very Big Brother feel to it."[82] For several weeks following the system's installation, opponents organized street protests in Ybor City, where demonstrators wore gas masks, Groucho Marx glasses, and bar code stickers on their foreheads. Organizers included the ACLU of Florida as well as the Ybor Coalition and the Tampa Bay Action Group. The *New York Times* reported that one protestor "walked by a camera, gestured obscenely and shouted, 'Digitize this!'" and *USA Today* reported that another protestor wore a computer monitor with Mayor Dick Greco's face on the screen.[83] References to "spy cameras," "digital lineups," and "facial frisking" circulated in the press coverage, registering specific anxieties about facial recognition technology. A reporter from *U.S. News and World Report* called the Ybor City experiment a "real-life version of *The Truman Show*."[84]

The local and national attention the project received prompted a response from local Tampa officials. Some members of the Tampa City Council began to question publicly whether the project should have gone forward, and there was some indication that several council members had not fully understood what it was they had approved. On July 4, 2001, the *St. Petersburg Times* reported that "snoopy software has council foe," referring to a statement made the day before by Tampa City Council member Rose Ferlita. Ferlita supported the original CCTV system, but opposed the use of facial recognition technology: "I haven't changed my position . . . I still feel the same way about regular, everyday surveillance cameras. But this next level is overpowering."[85] The addition of facial recognition technology seemed to Ferlita and others to signal something more insidious than standard video surveillance. Several Tampa City Council members joined Ferlita in opposing the project, while a few others came out publicly in support of it. In a move that in retrospect appears an exercise in public appeasement, the City Council considered a motion to advocate terminating the city's contract with Visionics. If passed, the motion would serve as a recommendation to the Mayor's Office, which had the final word about whether the project would proceed. Before the City Council meetings, however, the mayor publicly expressed his support for the project and his intention to allow it to go forward regardless of what the City Council decided. "As far as I'm concerned, the cameras will stay, no matter how the city council votes," insisted Greco.[86] The mayor dismissed objectors as naïve luddites: "The people who are opposed to this just don't understand it. . . . When I was mayor the first time, we were the first police department to use a helicopter, and people were against that. They thought we were going to spy on them from the air. Now every major department has one."[87] Ultimately, Greco did not have to go against the council's recommendation. After an initial tie vote as a result of an absent member, the motion to recommend termination of the project was defeated at a second council meeting by a vote of 4–3. Chairman Charlie Miranda cast the deciding vote, justifying his position with reference to the ubiquity of surveillance cameras in general and the irrelevance of the City Council decision.

Supporters of the project defended it on a number of grounds. The Tampa Police spokespeople dismissed the issue of privacy, repeating the legal refrain that people have no "reasonable expectation" of privacy in public. Police and other supporters also made the argument that the technology was more or less the same as standard police practice, only facial recognition technology would be more effective, faster, and more accurate than human police officers alone. According to Detective Todd, "This is no different than a police officer standing on a street corner with a handful of pictures, except for that it's more accurate and stops trouble faster."[88] Todd also suggested that facial recognition technology was a laborsaving device and a police force multiplier, allowing the TPD to "maximize the process of pointing out people we're looking for without putting 20 more officers on the street."[89] Tampa City Councilman Bob Buckhorn, who had shepherded the original proposal to install the system through the approval process, likewise became an outspoken proponent of the technology, defining it primarily in terms of its similarity to standard police procedures and its laborsaving benefits:

> I think what we are doing with facial recognition technology is merely applying modern technology to age-old policing techniques. When a police officer goes to roll call in the morning, he's given what's called a hot sheet [picks up a piece of paper to demonstrate]. A hot sheet contains a list of wanted individuals. He walks around with that in his hand all day long, keeps it in his car. What we are doing is just merely dumping a database

> of known offenders, of wanted individuals, sexual predators, lost children, into what is a modern means of identifying people. So to me it's no different than what the beat cop used to do, which would be walk around with that hot sheet. We're just using technology to do it in a more sophisticated, less expensive, less time-consuming fashion.[90]

In advocating the police use of the technology, Councilman Buckhorn wanted to reassure people that it was nothing radically new, just an augmented and more efficient form of identification, less costly, more high-tech, and faster than human police officers. His justification aimed to marry a nostalgic notion of "age-old" policing to the technological project, appealing to a perceived desire for a simpler, lost moment of cops-on-the-beat, while at the same time making a claim about the superiority of the technology over earlier forms of police practice. Buckhorn's defense of the project also suggested that the composition of the watchlist database was a settled matter, and that it contained records only of those in clear need of police apprehension. Although there was no explicit policy about "sexual predators" or "lost children," these figures became the preferred targets of the system among its defenders, repeatedly paraded out to establish the critical need for Smart CCTV. Visionics CEO Joseph Atick, for example, reiterated the frightful image of the sexual predator: "Wouldn't you want to know if a murderer or a rapist is sitting next to you while you're eating a sandwich? I would."[91]

The technology also had supporters among local Tampa residents, who maintained that police use of Smart CCTV was warranted in order to protect their right to security in Ybor City. In a letter to the editor in the *Tampa Tribune*, Patricia A. Benton, resident of the Tampa suburb of Seffner, expressed this sentiment:

> I will not go to Ybor City at any time, day or night, nor will I take out-of-town guests to visit there, because of the crime. To go there is to ask for trouble. Too bad that a person cannot visit the shops and restaurants anymore without fear of being carjacked, raped, or killed. And now we have a modern invention that will curtail that activity. But wait! It may infringe on our precious "rights." I have rights, too. I have the right to go where I please in public without worrying about being harmed. And the police have the right to utilize modern inventions that will secure that end. The framers of the Constitution would hide their heads in shame to know what we have come to, when the rights of criminals are more protected than the rights of honest citizens.[92]

It is not difficult to read Ms. Benton's expressed fear of crime as a salient problem in itself, regardless of whether she ever had been or would be a victim of crime. In fact, Ms. Benton saw herself as *already victimized* by the criminal class that threatened her freedom to shop, visit restaurants, and "go where she pleases." Of course, the local media's preoccupation with crime may have given Patricia Benton reason to fear being carjacked, raped, or killed in Ybor City, and overblown descriptions of a fully functioning facial recognition system encouraged the view that the technology could help "curtail that activity." For Benton and others like her, the new surveillance technology offered a legitimate and necessary means of police protection, violating the rights only of those who do not deserve them. This line of argument reiterated the comments of a politician responsible for the Newham facial recognition project in London, in response to objections from privacy advocates: "Yes, it is a civil liberties issue," a Newham councilman noted. "Our priority is the liberty of the people of this borough to go about their business without fear of crime. The rights of the majority are the most important consideration, not the rights of criminals."[93]

Pitting of the rights of "the majority" against an essentialized class of criminals is a stark dichotomy at the center of the crisis of penal systems and the corresponding transformations in policing that have taken place in Britain and the United States since the 1970s.[94] The expressions of Patricia Benton, concerned citizen, fueled a discourse of crime policy that "consistently invokes an angry public, tired of living in fear, demanding strong measures of punishment and protection."[95] In experimenting with new surveillance technologies, the Tampa Police were not simply imposing a vision of high-tech crime control on an unwelcoming public, but were in many ways responding to the demands of preferred groups for protection and secure access to public spaces. One can hardly fault Ms. Benton for wanting to move about in public spaces without being attacked. But her claim to the city expressed a sense of entitlement to public space that, far from holding out a vision of open access for all, was infused with resentment and a contentious politics of exclusion. As Doreen Massey has argued, "notions of place as source of belonging, identity, and security" are deeply tied to "notions of the self in opposition to the other that threatens one's very being."[96] Again, the claim that facial recognition technology targeted only specific, dangerous identities glossed over the more troubling politics of inclusion and exclusion that informed the redesign of urban space and its accompanying technological infrastructures. The aim of the Smart CCTV system was to provide a certain class of individuals with a sense of security in Ybor City, the sense of security of Patricia Benton and others like her from the Other that ostensibly threatens their very being. Much like the "community policing" paradigm that Christopher Wilson examines, technological projects designed to create that sense of security are not aimed exclusively at maintaining order, but also at reestablishing the legitimacy of police to decide "which communities are in a community and which are not."[97]

"Protecting the public has become the dominant theme of penal policy," argues Garland, and "the risk of unrestrained state authorities, of arbitrary power and the violation of civil liberties seem no longer to figure so prominently in public concern."[98] In the case Ybor City, civil liberties did have some resonance in public discourse about police adoption of the new surveillance technology, as we have seen. In fact, the conflict and controversy over the Smart CCTV project underscores a long-standing tension inherent in liberal governance between "the twin dangers of governing too much . . . and governing too little."[99] Liberalism denotes a certain ethos of governing that must constantly strike a balance between these two poles, writes Nikolas Rose. Governing too much means threatening to distort or destroy "the operation of the natural laws of those zones upon which good government depends—families, markets, society, personal autonomy and responsibility."[100] Governing too little means "failing to establish the conditions of civility, order, productivity and national well-being which make limited government possible."[101] The effort to integrate automated facial recognition with CCTV for the mediated supervision of Ybor City was a project caught up in this tension, and whether and how it would be made to work as a functioning technology would depend on whether the acceptable balance could be negotiated—and especially whether people were convinced that more sophisticated police surveillance technologies were a necessary prerequisite to their "freedom."

## *Drawing a Blank*

On September 11, 2001, events intervened to change the terms of the debate over police use of facial recognition technology in Ybor City, at least temporarily. The 9/11 terrorist attacks, coming just three months after the Ybor City Smart CCTV experiment began, instigated a barrage of press and policy attention to facial recognition and other biometric technologies. Companies marketing the technologies moved to capitalize on the heightened climate of fear and anxiety, positioning their products as among the range of requisite "security solutions." If most federal policy makers had barely paid attention to the Ybor City experiment with "smart" surveillance before, it now seemed to require urgent attention. The congressional committee hearings that Rep. Dick Armey requested about the use of facial recognition technology for public surveillance did in fact take place, three months after his request, in October 2001; however, the deliberations were not about the appropriate scope and limitations of police use of new "smart" surveillance technologies, but rather how quickly such technologies could be deployed at airports, border control stations, and other sites.

Still, the momentum given to the project in the aftermath of 9/11 did not force opponents of Smart CCTV to acquiesce to the use of facial recognition technology in Ybor City. In January 2002, the ACLU renewed their challenge to the project, releasing a report titled *Drawing a Blank: The Failure of Facial Recognition Technology in Tampa, Florida*. In the report, the ACLU made the case that facial recognition technology simply did not work and so represented a misdirection of security priorities. It referred to federal government tests (the FRVT 2000) where even the best products performed only moderately well in controlled laboratory settings. It also provided evidence, from documents received under Freedom of Information requests, revealing that the Tampa Police stopped using the facial recognition system less than two months after they began using it, precisely because of its poor performance.[102] The report had an undeniably negative impact on the Smart CCTV project, but it did not put an immediate end to the experiment. Shortly after the report was released, Visionics announced that the system was being upgraded to run on more than one video, grabbing faces from six video feeds simultaneously and thereby reducing the operator's need to switch cameras at his or her discretion. Still, no facial identifications materialized, and it remains unclear whether the Tampa Police began using the facial recognition system again in earnest. Press coverage of the project waned, and it received little or no public attention for over a year.

Then, in August 2003, the police experiment with facial recognition technology again made headlines: "Ybor cameras won't seek what they never found," declared the *St. Petersburg Times*.[103] In one last move, the Tampa Police issued a public statement announcing their termination of their contract with the company, then called Identix. The *Tampa Tribune* reported that the system was shut down on Tuesday, August 19, "having failed in its objective" to recognize "the facial characteristics of felons and runaway children."[104] According to Tampa Police Captain Bob Guidara, the facial recognition system "was of no benefit to us, and it served no real purpose."[105] Others spun the termination of the project differently. Tampa Police spokesman Joe

Durkin said he "wouldn't consider it a failure. . . . You are always looking for new and efficient ways to provide the best service to the community. There's going to be ups and downs."[106] Identix offered a one-sentence statement that defended the company as a responsible corporate citizen and suggested that the public mood was not right for the system's implementation: "Identix has always stated that this technology requires safeguards, and that as a society we need to be comfortable with its use."[107] But Joe Durkin insisted that police discontinued using the system "because of the lack of arrests, not the privacy issues."[108]

## *Smart CCTV or No Smart CCTV?*

The controversy over the Ybor City Smart CCTV experiment was, fundamentally, a struggle over the appropriate extent and limitations of police power, a balancing act that has consistently posed a challenge to liberal democratic societies and one that seems to lean, in a neoliberal climate, toward a prevailing belief in expanding police power as both necessary and inevitable. It would be wrong to assume that the initial installation of the police CCTV system in 1997 was itself universally accepted, but by the time the Smart CCTV project began in 2001, many people had more or less accepted the idea of video surveillance in public spaces. It was the idea of automated facial recognition in particular that generated conflict, invoking in the public imagination competing visions of a brave new technological future. While some nostalgically hoped for the return of a recovered, crime-free community from the mythic past, others saw an urban dystopia in the frightening mold of *1984*, a prison-like environment devoid of all freedoms where everyone is under the constant gaze of the police. For opponents, the police experiment with facial recognition technology in Ybor City demonstrated a police power grab over and above the use of "basic" video surveillance, essentially turning every person on the street into a criminal suspect. But while the move to upgrade the CCTV system gave opponents an opportunity to reignite the debate over police surveillance in Ybor City, it is important to recognize that shutting down the CCTV system itself was never considered as a viable option (which is not to say that no one raised the issue). The facial recognition project may have been abandoned, but the CCTV system continues to generate images designed to keep Ybor City under constant, mediated police supervision.

Ultimately, the effort to integrate facial recognition technology with video surveillance failed for contradictory reasons. The project did indeed suffer to some extent as a result of the successful efforts of vocal opponents to define automated facial recognition as a technology that gives the police too much power. On the other hand, there were others, including the police themselves, who viewed it as an ineffective technology of crime control because it never managed to identify anyone successfully. Of course, as some have pointed out, the lack of positive identifications may have been evidence that the system was serving as an effective deterrent, keeping wanted criminals away from Ybor City.[109] But since addressing the *fear* of crime was as important as actually preventing it, the police needed a success *story* in order to sell the Smart CCTV system: a narrative of a vile criminal identity—preferably a rapist, murderer, or child molester—being apprehended thanks to the facial recognition system. For reasons that extend beyond the specific technical limitations of facial recognition technology, the police never acquired the material they needed to "create" such a story (and fortunately they did not have as much latitude as the production team at *American Detective*). Without such a story, or multiple stories, Smart CCTV became more of a liability than a benefit for the Tampa Police, denying them the glory of identifying and catching the bad guys, and leaving them only with the perception of a power grab based on a faulty technology. In short, it offered the police neither an immediate practical solution to the "video burden" nor a compelling symbolic display of their technological crime-fighting sophistication.

The experiment with facial recognition technology in Ybor City ended without the permanent adoption of Smart CCTV by the Tampa Police, but spokesman Joe Durkin was probably correct to qualify the term "failure." The negative attention the project received throughout the process from the ACLU and other parties made it impossible for the developers to define the technology on their own terms, and the termination of the project could not help but set back efforts to portray automated facial recognition as a viable technology. But from the beginning, people directly involved in the project understood the highly experimental nature of what they were doing, and despite public statements about a smoothly functioning system, they were likely well aware that there was no guarantee the experiment would be successful. To make facial recognition technology work with video surveillance systems in urban spaces, it must be tested and developed in those spaces, and only through a series of "ups and downs," advances and setbacks, will the necessary improvements be made that transform Smart CCTV from a set of experiments to a functioning police technology. As long as the diffusion of CCTV systems proceeds apace, fueled by the normalization of crime and the persistent pressure on the police to appear in control of "the crime problem," then the experimentation with new technologies for optimizing CCTV functionality will likewise carry on. Rethinking this largely ineffective approach will require a full-scale effort at redefining the problem—another kind of legitimation campaign aimed at defining crime not as a cause, but rather an effect of social disorder; not as a normal part of everyday life and a forgone conclusion for certain kinds of people, but rather a product of deepening social inequalities tied to structural conditions. Without this process of redefinition, we will witness not only the persistent police pursuit of more sophisticated and effective surveillance technologies, but also the construction of many more prisons and walls behind which to consign the expanding disenfranchised and individuated criminal classes.

# 3

# Finding the Face of Terror in Data

During the Cold War, the enemy was predictable, identifiable, and consistent. We knew the threats, the targets were clear. But times change. Today, with the demise of the other superpower, America is in a different position: a position of vulnerability. When the enemy strikes, it isn't predictable. It isn't identifiable. It is anything but consistent. Times change. We are in a world of "asymmetrics," and we need transformational solutions. The asymmetric threat is now a reality of global life. How do we detect it? How do we predict it? How do we prevent it?

—Promotional video for the Total Information Awareness program, 2002

In a top ten list for 2000, the next nine countries' defense budgets do not *add up* to the United States's. As remarkable as the sheer size of the military budget might be, it begs a larger question, which in the rush to reach a budget agreement went mostly undebated: just where is this enemy who justifies such expenditure?

—James Der Derian, *Virtuous War*

## *From the Cold War to the War on Terror*

In his keynote address at the September 2002 Biometrics Consortium Conference, Dr. Robert L. Popp, then deputy director of DARPA's newly formed Information Awareness Office, began by showing a promotional video for the "Total Information Awareness" (TIA) program. TIA would later spark public controversy and bipartisan political opposition, but the press had not yet taken notice of it, and the conference audience of industry and government officials seemed interested and receptive. The video, outlining the various research and development projects combined to form TIA, opened with a montage of images and sounds signifying the Cold War, the fall of the Berlin Wall, and the newly defined U.S. security threats of the 1990s and early twenty-first century. The Cold War images—black and white photos of suffering peasants and video of Soviet soldiers marching in file—were followed by mug shot images of recognizable terrorist suspects, including Osama bin Laden, and a familiar video image of a crowd of Arab men moving rhythmically en masse. This visual montage accompanied the voice-over narration quoted above proclaiming "the asymmetric threat" as "now a reality of global life."

Providing a simple narrative of transition from the Cold War to the "war on terror," the video contrasted these new "asymmetric threats" with the more symmetrical geopolitical conditions that existed prior to the breakup of the Soviet Union. The text invoked a nostalgic longing for the predictability and consistency of the Cold War, when the enemy was ostensibly well-defined and *identifiable*, combining it with an almost gleeful futurism about the promise of new digital technologies to save the West from its uncivilized Other. The idea of an "unidentifiable" enemy presented a new form of national vulnerability—"America" was now "vulnerable" precisely because it could not identify its enemy. The United States seemed more in danger than ever, with the "asymmetric threats" facing the nation in the post–Cold War context even greater than the perpetual threat of nuclear holocaust during the arms race with the Soviet Union. The collapse of the communist regimes may have dissolved the symmetric threat of nuclear warfare with the "other superpower," but in its place came many other villains more difficult to locate, define, and identify.

Although it is not difficult to discern problems with the suggestion that nation's Cold War enemies were easily identifiable, the fall of communism did in fact create new security risks for the United States, destroying the balance of global power that had created a relatively stable international order in the four decades following World War II.[1] In addition to specific new threats, the demise of the "other superpower" presented another sort of vulnerability for the U.S. national security state: a crisis of legitimacy. In the 1990s, a rising volume of criticism questioned why the United States was still spending ten times as much on defense as its closest defense-spending competitor, and nearly as much as the rest of the world combined. Just where was the enemy that justified this expenditure? As one response to this question, military strategists placed special emphasis on "asymmetric threats," a trope that not only embodied a questionable claim to the unique nature of the new post–Cold War adversaries, but also invested relatively small threats with greater threat potential, aiming to provide some justification for the ongoing reproduction of the imperial-sized, Cold War national security apparatus. The United States may no longer have an enemy that could match its military might, according to this message, but it now has more insidious enemies that do not play by the conventional rules of state warfare, and thus represent significant threats to the nation, disproportionate to their relatively minuscule military resources.

The military discourse that defined these new "unidentifiable" and "inconsistent" enemies as major security threats was given considerable leverage by the enormity of violence on 9/11 along with its precession as simulacra. The TIA promotional video exemplified the way that state security discourse constructed both the threat of terrorism and the appropriate security strategies to address that threat in the aftermath of the catastrophe. The ambiguous signifier of asymmetric threats became a chosen mantra of state security discourse, an allegedly new type of danger that required considerable research and investment into new types of war-fighting technologies. The tropes of "asymmetric threats" and "unidentifiable enemies" provided justification for a set of policy proposals after 9/11, including arguments in favor of stepped-up funding for the development and widespread deployment of facial recognition and other biometric identification systems. Given the problem of small but highly dangerous enemies that were increasingly difficult to identify, the need to develop and deploy expensive, new, high-tech surveillance technologies at a proliferation of sites seemed self-evident.

There were of course other forces at work shaping the political response to 9/11. As we saw in chapter 1, the political-economic priorities of neoliberalism had a major influence on the demand for network security technologies in the 1980s and 1990s, including facial recognition and other biometrics, and these priorities played a significant role in defining the political response to the attacks. It was immediately apparent that the events of 9/11 would be a major boon to providers of security systems and services, an industry deeply connected to the information technology sector. As the enthusiastic response of both security and IT brokers clearly evidenced, post-9/11 security provision would involve ventures aimed at maximum profitability, and the business of security would overlap considerably with the business of information technology. In fact, long before 9/11, the information and security sectors were so tightly integrated as to be virtually inseparable; the major players in the IT sector were hard at work developing network security systems, and both state and private-sector entities conventionally understood as security providers had long since integrated IT into all manner of security systems. Thus it was unsurprising to find an industry observer from *Intelligent Enterprise* noting with optimism that "homeland security spending will help fuel an IT recovery. IT solution providers may some day look back on the War on Terror and be grateful for the opportunities born out of turmoil."[2] Such pronouncements not only articulated the overlapping dimensions and priorities of security and IT, they also offered a clear expression of the market logic that would define what "homeland security" and the "war on terror" meant in practice.

Proponents of biometrics in particular saw a major opportunity to capitalize on the emerging "homeland security" regime, and in fact to participate in the very construction of the strategies and programs that would define what homeland security looked like in practice. The 9/11 attacks happened at a time when vendors of facial recognition systems were beginning to experiment with real-world applications, from Department of Motor Vehicle offices in the United States, to voter registration systems in Mexico and other countries, to Smart CCTV experiments in London and Tampa. As we saw in chapter 1, the application of biometrics for the "securitization of identity" was already taking place to satisfy industry demand for priorities like network security, labor control, and consumer tracking. State agencies likewise had already begun adopting biometrics as part of the more repressive apparatuses of the penal-welfare-immigration-control complex. In the language of cultural studies, the aftermath of 9/11 was a moment of articulation, where objects or events that have no necessary connection come together and a new discursive formation is established: automated facial recognition as a homeland security technology, a means of automatically identifying the faces of "terrorists." The interests of biometrics industry brokers to push their technologies after 9/11 translated well into the prevailing public policy and press response to the attacks: the frenzied turn to "security experts" to speculate as to the source of security failures and to provide recommendations for "stopping the next one."[3]

The biometrics industry readily answered the call for expert knowledge of security issues and technologies, positing their identification systems as *the* solution to the new terrorist threat. The spokespeople for the biometrics industry worked feverishly to promote biometrics as central components in new security systems and to situate themselves and their companies as "moral entrepreneurs" taking charge in a moment of national crisis. Industry brokers issued press releases, appeared in the press on a regular basis, and testified before Congress as to the benefits of their products. Most impudent, proponents of facial recognition technology repeatedly suggested that such systems could have prevented at least one if not all of the hijackings.

While military needs influenced the development of automated facial recognition since its inception as a research program, the technology took on special importance in state security discourse after 9/11, as advocates took the opportunity to catapult it from set of technological experiments into something more closely resembling what Bruno Latour calls a "black box"—a functioning technology positioned as virtually indispensable to a secure, technological future.[4] Facial recognition technology would ostensibly provide an accurate, objective, precision-guided means of identifying the faces of terrorists as they attempted to pass through airport security, border control stations, and a proliferation of other checkpoints. The technology promised to succeed where human security staffers failed, compensating for their imperfect, subjective perceptual abilities and limited memory capacities.

This chapter examines the preoccupation with facial recognition technology in the post-9/11 context, unpacking claims to its technical neutrality by investigating the cultural logic that defined the technology and the practical politics that shaped system development. While the promise of facial recognition lay in its potential to individualize the terrorist threat by targeting specifically identified "terrorist" individuals, the effort to define it as a homeland security technology also made use of an implicit classifying logic, including rhetorical moves that resuscitated antiquated notions of deviant facial types. In practice, facial recognition technology seemed uniquely suited to identifying the individual faces of "terrorists." But in a more symbolic sense, the technology promised to provide a means of "protecting civilization" from a more generalized and racialized "face of terror." Although facial recognition algorithms were not expressly designed to classify faces according to racial typologies, the symbolic authority of the technology in the post-9/11 context very much depended on the idea that it could in fact be used to identify a mythic class of demonic faces that had penetrated the national territory and the national imagination. The "facialization" of terrorism—defining non-state forms of political violence with recourse to the racist logic of a mythic, demonized facial type—was prevalent in discourse about facial recognition technology, appearing alongside claims about its technical neutrality.

If claims to the technical neutrality of automated facial recognition disavowed the meanings assigned to it, they also denied the forms of social differentiation it promised to accomplish, offering too narrow a view of how surveillance and identification systems function. The application of automated facial recognition for "terrorist" identification required more than the development of algorithms for digitizing the face or the deployment of biometric capture devices. It also required the construction of a terrorist classification system—a technology for *making up terrorists*—that took the tangible form of a consolidated watchlist database. Like the central role of the archive in the application of photography to criminal identification, the database is at the heart of biometric system development. In order to understand how facial recognition systems function, it is crucial to have some sense of how facial image databases are constructed, especially the social forces governing their form and content. How are facial image databases populated? What identities are entered into a database and why? What information about them is archived? How is that information organized, and how is the "accuracy" of the information determined? As noted in the previous chapter, these are policy questions, as well as questions about the practical design of identification systems. But they are also questions about the meaning of facial recognition technology, and about the politics that inform system development. A close look at terrorist watchlist database consolidation demonstrates the way that classification systems, enshrined in the material form of the database, construct rather than merely reflect the identities they purportedly represent. It also shows that, in practice, there is nothing neutral about the way computers are being programmed to "see" the face.[5]

## *Racializing the Terrorist Threat*

On the eve of the second anniversary of 9/11, the *New York Times* published an op-ed piece by John Poindexter, the former national security adviser to Ronald Reagan best known for his involvement in the Iran-Contra scandal. Poindexter had recently been appointed as head of the Total Information Awareness program, a set of funding initiatives for research and development into new data mining and analysis technologies that would make optimal use of the full range of available public- and private-sector databases to gain knowledge about the identities and activities of "terrorists." Although all the research programs gathered together under TIA existed in some form before its creation, in its newly organized form it received widespread criticism as having an overly Orwellian mission to spy on Americans. As a result of bipartisan opposition to TIA, Congress moved to defund the program in the summer of 2003. Poindexter's op-ed piece in the *Times* was an effort to defend TIA's parent agency, DARPA, from further funding cuts, arguing for the importance of DARPA's research programs and the agency's neutrality with respect to any applications that resulted from the research it funded.

If Poindexter's appeal itself was important for what it revealed about the politics of military R & D in the post-9/11 context, it was less compelling than the headline of the article and the image it conjured: "Finding the Face of Terror in Data."[6] Referencing sophisticated new techniques of data mining, the headline also carried with it powerful connotations of national contamination, along with the implication that new digital technologies could be used to purify the nation of its enemies within. Computerized forms of data analysis and retrieval were continuously held out in the wake of 9/11 as a means of identifying hidden information vital to uncovering terrorist plots. The idea that Poindexter or DARPA or anyone else could "find the face of terror in data" implied that there actually existed a "face of terror," a particular type of face characteristic of terrorists, and that large data sets could be mined in search of relationships and patterns that would reveal their identities. In other words, in the process of infiltrating civilized society, "terrorists" have left traces of their presence in the parallel world of data, society's digital mirror image. Like Trojan horse computer viruses, they would need to be identified, isolated, and eradicated using the most sophisticated forms of data mining and analysis. The technological projects cobbled together under DARPA's TIA program each claimed to provide a technical means of national purification, a sophisticated, high-tech approach to targeting external and internal threats to the population so that the nation could remain healthy and vibrant in the face of newly recognized hazards and contaminants.

Neither Poindexter's headline nor the article itself made explicit claims about facial recognition technology specifically, but it required no stretch of the imagination to see it there, as one among an array of technical means that would ostensibly help authorities locate the "face of terror" in the field of data circulating over global IT networks. In fact, one of the programs placed under the TIA umbrella during the program's brief tenure was the "Human ID at a Distance" initiative, or Human ID for short, a DARPA program formed after the bombing of the Khobar Towers U.S. military barracks in Saudi Arabia in 1996. Originally called "Image Understanding Force Protection," Human ID provided funding to university research labs and private companies doing research on biometric technologies, with the ultimate aim of developing and fusing multiple biometrics (face, voice, gait) into one, more robust automated sensory system that could identify individuals from a distance, such as around the perimeter of U.S. military installations. The goal was to develop image analysis technologies that could be applied to improve on military surveillance systems and protect U.S. military forces abroad, especially in the Middle East.

DARPA's Human ID biometrics funding initiative was part of a broader U.S. military strategy, no less than a "revolution in military affairs" aimed at developing command, control, and communications technologies in order to achieve "global information dominance."[7] The history shared by computer science and engineering on the one hand, and military research and development on the other, has been well documented.[8] The drive to develop autonomous weapons systems in particular has been at the center of artificial intelligence research since its inception, including the effort to invest computers with synthetic forms of perception (vision and hearing).[9] Although the military has never fully handed decision-making authority over to computers, military applications of AI have increasingly blurred the distinction between merely "advisory" and fully "executive" capabilities.[10] The development of imaging technologies that could automatically identify targets "at a distance" has been a part of this effort to create new and more effective human-computer war-fighting assemblages, with a certain level of authority and perceptual capacity translated into automated systems. Fueled by a strategy of "global vision," military R & D has aimed to integrate new visualizing technologies and weapons systems in order to form what Paul Virilio has called a "logistics of military perception," whereby the major global powers dominate using technologies for attending perpetually and immediately to images and data, keeping targets constantly in sight.[11] Under these conditions, according to Virilio, the perspective of "real time" supersedes the perspective of "real space" (invented by the Italians in the fifteenth century), and seeing over distance becomes virtually synonymous with contact—and killing—over distance. The war-fighting promise of automated facial recognition and related technologies lay precisely in their potential to identify objects automatically and "at a distance," whether the final aim was to control, capture, or destroy these targets.[12] In short, in the military context the aim of identification-at-a-distance has been inescapably married to the tactic of killing-at-a-distance.

Political theorists have debated at some length about the paradoxical nature of the modern state's sovereign claim to the right to kill. For Foucault, this problem became especially salient with the emergence of political systems centered on the health and well-being of the population—in other words, with the rise of what he calls "biopower."[13] The historical emergence of this form of political power corresponds to the discovery, along with the formation of the sciences of demography and statistics, of a new object of state intervention in the eighteenth and nineteenth centuries: *the population*. The life of the population itself, as an aggregate of bodies in circulation with one another and with their social and environmental conditions, became an object of intense political and scientific concern. The political technology of biopower became embedded in existing disciplinary techniques that targeted the individual body, but took a broader view of the individual as a component of a larger collective body, one that had its own qualities and required its own regulatory interventions. Along with the discovery of the population came the corresponding recognition of its variables and vulnerabilities—the forces, such as disease epidemics, draughts, crimes, suicides, and procreative and child-rearing practices, that affected the health of the population. These forces in turn needed to be measured, evaluated, and intervened on in a manner that would make the population as a whole healthier, more secure, and more productive. A whole set of institutions and bodies of knowledge emerged to examine and deal with the health and security of the population, including medicine and public hygiene, insurance, individual and collective savings, safety regimes, and eugenics and antimiscegenation campaigns as well as other formal and informal regulatory controls on sexuality and reproduction.[14] These bodies of knowledge were not strictly repressive but likewise aimed to productively define the range of active, normal, civilized forms of human subjectivity that would be appropriate for modern citizens living in "free" societies.

The question arose, under emerging forms of biopolitical governance, as to how such political systems concerned centrally with the health, well-being, and security of their populations could also claim the sovereign power to kill, "to call for deaths, to demand deaths, to give the order to kill, and to expose not only [their] enemies but [their] own citizens to the risk of death."[15] It is at this point, Foucault argues, that racism intervenes. Racism becomes "a way of introducing a break into the domain of life that is under power's control: the break between what must live and what must die."[16] Racism enables the biopolitical differentiation of the population into categories of relative value in the name of the health and security of the population as a whole. When the security of the total population is paramount, racism becomes the means by which the state claims the legitimate right to kill, and especially to enable its citizens to be killed. To be clear, by "killing" Foucault does not simply mean murder as such, "but also every form of indirect murder: the fact of exposing someone to death, increasing the risk of death for some people, or . . . political death, expulsion, rejection, and so on."[17] The modern state "can scarcely function without becoming involved with racism at some point."[18]

The biopolitics and associated forms of state racism that emerged in the eighteenth and nineteenth centuries are still with us today but continue to evolve and take new forms. "Racism does not . . . move tidily and unchanged through history," writes Paul Gilroy. "It assumes new forms and articulates new antagonisms in different situations."[19] Likewise, new conditions and instabilities have arisen that challenge the health of populations in large-scale societies, providing impetus for new forms of biopolitical intervention.[20] Many of the global instabilities and crises that the populations of modern nation-states face today are without a doubt what Anthony Giddens calls "manufactured risks"—that is, risks resulting not from natural forces external to human activity, but from human development itself, especially from the progression of science and technology.[21] International terrorism is a paradigmatic example of a "manufactured risk," arguably arising, in the most general analysis, as a result of diverse and even incommensurate cultures coming in contact thanks to imperialistic and expansionist impulses, along with associated developments in communications and transportation systems. Of course, this is not the explanatory picture that is painted of terrorist acts when they occur. Rather, they become acts of sheer and inexplicable evil, and they provide perfect fodder for the intervention of racism into biopolitical forms of government. This was especially salient in the post-9/11 context. As images of Osama bin Laden and the alleged hijackers circulated in the Western media, a powerful metaphoric image of the "enemy Other" took shape. The racialization of the enemy was virtually assured, providing a ready alibi for intensified state investment in the technological infrastructure to support the biopolitical differentiation of the population. Despite the language of "unidentifiable enemies," finding the face of terror in data meant designing and deploying what Lisa Nakamura calls the "socioalgorithmics of race"—new technologies designed to target specific types of faces and bodies (those of Arab men, to be sure, but also all manner of other derelict identities).[22]

The practice of constructing a racialized image of a mythic enemy has long functioned as a way of solidifying and reinforcing national identity.[23] At least since World War II, propagandists have recognized that "the job of turning civilians into soldiers" could be achieved through the uniquely effective tactic of superimposing a variety of dehumanizing faces over the enemy "to allow him to be killed without guilt."[24] The trope of the "face of terror" that circulated widely after 9/11 functioned in this way, offering a caricatured version of "the enemy," while at the same time suggesting the existence of a terrorist facial type. Along with ubiquitous images of the faces of Osama bin Laden and the hijackers—the alleged unidentifiable enemies—the "face of terror" invoked specific objects: mug shots and grainy video images of Arab men. The trope resuscitated a history of creating archetypical racialized enemies, along with the associated practice of "facialization," Karen Engle's term for the process whereby "the face of a subject, which is believed to reveal an interior truth kernel or deep essence, comes to stand for the narratives a nation tells about itself."[25]

It was not John Poindexter but Visionics Corporation that was the first to make use of the "face of terror" trope. In its initial effort to position itself at the center of the public-private security response to 9/11, the company released a policy "white paper" on September 24, 2001, titled *Protecting Civilization from the Faces of Terror: A Primer on the Role Facial Recognition Technology Can Play in Enhancing Airport Security*. The bold-faced claim that the technology could "protect civilization" can be read as hyperbole only in retrospect; in the immediate aftermath of the attacks it represented a serious statement about the product's capabilities. The "faces of terror" phrase, obviously used as clever means of positioning facial recognition technology as a solution to airport security, also must be understood in the grave climate of the moment. While ostensibly referencing the individual faces of the 9/11 hijackers as well as potential future terrorists, it had more general connotations as well, signifying a metaphoric, racialized enemy Other—a demonic type of face that had penetrated both the national territory and the national imagination.

Other uses of the "face of terror" trope in association with facial recognition technology followed. For example, the Technology, Terrorism, and Government Information Subcommittee of the U.S. Senate Judiciary Committee held a hearing on "Biometric Identifiers and the Modern Face of Terror: New Technologies in the Global War on Terrorism."[26] Like Visionics' use of the metaphor, such references could be interpreted as merely clever turns of phrase if not for the seriousness of the moment and the extent to which they made unveiled claims about the existence of a terrorist facial type. The "face of terror" trope embodied both an individualizing and a classifying logic of facial representation, sliding from one meaning to another. Interpreted in correct grammatical terms, the phrase referred to individuals with expressions of terror on their faces, but it was used instead in reference to the perpetrators of terrorist acts. The "face of terrorism" would have made more sense but was likely too explicit in its reference to a terrorist facial type. (George W. Bush came closest to using this reference in a statement announcing the FBI's "most wanted terrorists" watchlist: "Terrorism has a face, and today we expose it for the world to see."[27]) In its reference to a representative terrorist face, the trope could not help but resuscitate assumptions from the antiquated science of physiognomy about the existence of relationships between facial features and individuals' essential qualities, including alleged inherent propensities for criminal or other deviant behaviors.

But the image of "the modern face of terror" promulgated after 9/11 did not take precisely the same form as its physiognomic predecessors. Repackaged in digital form and distributed over computer networks, it was more akin to a "virtual enemy"—the simulated adversary of what James Der Derian ironically calls the "virtuous war."[28] This type of war is built on a new war machine, "the military-industrial-media-entertainment network," which merges simulation technologies with lethal weaponry, action movies and video games with civilian and military training exercises, and computer vision with "the logistics of military perception."[29] Specially designed for killing at a distance, this new war machine is no less violent than earlier forms, but claims to use high-tech war-fighting technologies in the service of virtue. When fighting the virtuous war at a distance, it is especially easy to kill the virtual enemy without guilt. And "the more virtuous the intention," writes Der Derian, "the more we must virtualize the enemy, until all that is left as the last man is the criminalized demon."[30] In its metaphoric connection to automated facial recognition, the demonic face of terror functioned like Der Derian's virtual enemy, invoking the image of a digitally generated avatar or morphed composite of "terrorist faces."

The mythic image of a morphed terrorist facial avatar embodied in the post-9/11 "face of terror" trope can also been seen as the antithesis of the now famous computer-generated image of a fictional woman's face printed on the cover of *Time* magazine in the fall of 1993. Deemed the "New Face of America," the image was morphed together from the facial images of seven men and seven women of various ethnic and racial backgrounds, and was used to promote a special issue on "How Immigrants Are Shaping the World's First Multicultural Society." Feminist scholars found the *Time* cover full of familiar and problematic assumptions about race and gender. The notion of race and pure racial types remained deeply embedded in the technique of computer morphology, Evelyn Hammonds argued, and morphing was "at the center of an old debate about miscegenation and citizenship in the United States."[31] The way the fictitious female face conveniently substituted the bodiless practice of morphing for the flesh-and-blood reality of miscegenation similarly made Donna Haraway uncomfortable, particularly to the extent that it effaced a bloody history and promoted a false sense of unity and sameness.[32] According to Laurent Berlant, the "New Face of America" on the *Time* cover was "cast as an imaginary solution to the problems of immigration, multiculturalism, sexuality, gender, and (trans)national identity that haunt the U.S. present tense."[33] The morphed image was feminine, conventionally pretty, light skinned, and nonthreatening, preparing white America for the new multicultural citizenship norm.

Like *Time*'s fictitious multicultural citizen, the post-9/11 "face of terror" was a similar sort of fetishized object, but in reverse. "The modern face of terror" was a technologically constructed archetype, and one for which racial categories still deeply mattered despite the absence of overtly racist references. Where the "New Face of America" allegedly represented progress toward an assimilated ideal, the "face of terror" trope deeply negated those same ideals of integration. The face of terror became an imaginary target for directed attention and hatred, but one that was likewise aimed at preparing the United States mainstream for new citizenship norms, especially the intensified practices of surveillance and securitization. Skillfully glossing over the tension between the individualizing and classifying logics of identification—"the tension between 'identity' as the *self-same*, in an individualizing, subjective sense, and 'identity' as *sameness with another*, in a classifying, objective sense"[34]—the "face of terror" trope helped to construct terrorism as a problem with a specific technological solution: computerized facial recognition.

LEFT: "The New Face of America," *Time* magazine's infamous morphed cover image "created by a computer from a mix of several races," November 18, 1993.
RIGHT: "TARGET: BIN LADEN," *Time* cover image close-up of Osama Bin Laden's face, October 1, 2001.

## *Security Through Intelligence-Based Identification*

Precisely how would facial recognition technology be used to identify the faces of "terrorists"? Proponents of the technology were well aware that it was not capable of identifying a "terrorist" facial type, and no credible authority ever made an explicit claim to that effect. In practice, facial recognition systems were being designed to identify individual faces, not classify facial types, an aim most developers would have recognized as a misguided technical goal. Instead, the technology promised to enable the faces of individuals designated as "terrorists" to be identified one by one. Individual faces captured by security cameras in airports and other areas would need to be automatically extracted from images, normalized to conform to a standard format, and digitized to produce mathematical "facial templates." But working systems would depend on more than the digitization of facial images or the installation of biometric capture devices. Significantly, identifying the faces of specific individuals designated as terrorists would also require amassing facial image databases to serve as the memory for automated identification systems. Every individual face to be identified would have to be included in a watchlist database. A massive and complex machinery of vision, built on an enormous image data-banking effort, would be required to make facial recognition systems function for "terrorist" identification.

The policy white paper that Visionics released two weeks after 9/11, *Protecting Civilization from the Faces of Terror*, offered a plan for integrating facial recognition technology into airport and other security systems, emphasizing the critical need for a major intelligence gathering effort to build the terrorist watchlist.[35] The document called for a large-scale initiative, "not only a drastic overhaul of the entire security infrastructure of airports, ports of entry and transportation centers in this country, but also around the world."[36] Under the heading "Security Through Intelligence-Based Identification," the authors outlined five applications for airport security and border control: facial screening and surveillance; automated biometric-based boarding; employee background screening; physical access control; and intelligence data mining. The white paper defined automated facial recognition as an already functioning technology, but one that required more "intelligence" in the form of knowledge of terrorist identities and photographs of their faces. Stopping short of providing the criteria for the designation of individuals as "terrorists," the authors enlisted "intelligence agencies around the world" to take responsibility for building these databases of terrorists' faces, which could then be used "to track them through computerized facial recognition as they travel from country to country." Once databases of terrorist faces were built from "covert photos and video footage supplied by operatives in the field," they could be networked together to provide an archive of terrorist identities for matching against facial images captured at airports and ports of entry to the United States. According to the company, the Visionics FaceIt system, then in the "deployment phase," would make it possible "to rapidly, and in an automated manner, use video feeds from an unlimited number of cameras and search all faces against databases created from various intelligence sources and formats," and then notify security or law enforcement agents in real time when the system located a match.[37] The watchlist database would be at the center of an effective, functioning facial recognition system for terrorist identification.

Long before the emergence of the database form, image classification systems, enshrined in physical archives, played a central role in the application of photography to the procedures of official identification. The role of the archive and archival practices gained importance with each innovation in photographic reproduction. As Allan Sekula has argued, in the application of photography to criminal identification systems in the nineteenth century, the camera alone was a limited technology.[38] The contribution of photography to identifying criminals came with its integration into a larger ensemble, "a bureaucratic-clerical-statistical system of 'intelligence.'"[39] The central artifact of criminal identification was not the camera, according to Sekula, but the filing cabinet. Sekula foregrounds the central role of the photographic archive as an apparatus of truth production in the Foucauldian sense. Archival systems organize and categorize the visual in ways that do not just reflect preexisting classification schemes but actually create them in the archiving process. Sekula describes the photographic archive as a "clearinghouse of meaning."[40] It "liberates" photographs from the context of their original use, offering them up for a greater number and variety of future uses. It allows new meanings to supplant the meanings derived from the original contexts in which photographs were taken. Yet archives also establish an order on their contents, operating according to an empiricist model of truth.

The electronic database form has inherited these characteristics and in many ways amplified them. The database differs from the conventional archive because it allows the user to access, sort, and reorganize hundreds, thousands, and sometimes millions of records, and it assumes multiple ways of indexing data.[41] The digital reproduction of information, the screen interface, software technology, and other features of the database form distinguish it from physical filing systems. And as a recent manifestation of the archival form, the database serves as "a new *metaphor* that we use to conceptualize individual and collective cultural memory," as Lev Manovich has argued.[42] The database functions as discourse, according to Mark Poster, a way of configuring language that constitutes subjects.[43] It is both an abstract paradigm and concrete mechanism for organizing information.

For most of its existence, the database form has been associated with the storage of text-based coded information, but databases are now increasingly used to house visual information as well. Thanks in part to the availability of cheap digital cameras, along with more computer storage and processing power, a veritable "avalanche of images" is flooding into databases. The image database is a cultural form of increasing significance in the organization of visual perception and visual meaning, and a technology that plays a central role in the effort to build computer vision systems. In both their form and content, image databases define the parameters of what computers can be programmed to "see."

Surveillance systems have long been designed to incorporate both visual and text-based forms of information; however, new techniques of digitization, database building, and image retrieval are being designed to provide more effective and automatic ways of processing and organizing visual media, including both still photographs and moving images (including an abundance of images generated from surveillance video). Oscar Gandy has used the term "panoptic sort" to describe a form of "high-tech, cybernetic triage" that uses databases of transaction-generated data to sort individuals according to their presumed economic or political value.[44] Gandy's use of the term "panoptic" is metaphorical—his analysis of the "panoptic sort" emphasized the use of text-based, coded information for the social sorting of people, rather than visual or optical information. But the "panoptic sort" is taking on a more literal, optical dimension along with the growth of image databases and the development of digital techniques of visual image processing. In the case of facial recognition systems, the aim is to translate images of the face into coded information and to automatically connect those coded facial images to text-based, documentary information about individuals. As noted in chapter 1, facial recognition technology promises to more thoroughly and effectively integrate visual and documentary modes of surveillance, which is precisely the innovation that makes it such a potentially powerful technology.

Given that database technology is becoming such an important part of the organization of visual information, how do we visualize the database? What does a database look like? Most likely what we visualize is the software interface, which most directly positions us as subjects relative to a database. Take for example a familiar image of a database-software interface: the screen image of the criminal or "Most Wanted" database. An updated version of the outlaw "Wanted" poster, the Most Wanted database serves an important symbolic function as a cultural image of law and order. The Most Wanted database is a common image in popular crime dramas, for example, where it tends to have fairly extraordinary search capabilities. Screen images that resemble the popular crime drama versions of the Wanted database can be found on the Internet, including one at the website for the television show *America's Most Wanted* (*AMW*) and one at the government website for the FBI.[45] The graphic design of the *AMW* website simulates a crime database interface. On the top of each page, visitors to the site see that their "Database Status" is "Online" and their "User Access" is "Approved." The production values of the *AMW* site surpass the FBI's "Most Wanted Terrorists" website, but otherwise the sites are similar in form. Both display rows and columns of mug shot photographs of wanted individuals, and each digital photo is itself a hyperlink to an individual record with additional information about the wanted person. Navigating these sites invites Internet users to envision themselves as intelligence agents, accessing information about wanted individuals in case of a personal encounter, or doing their patriotic duty to inform themselves about the nation's enemies.

The simulated versions of the Most Wanted database found at these promotional websites convey a particular image of how a watchlist database functions. As much through their design as their content, the sites appear to provide authoritative, unambiguous, empirical information about the

The FBI's ten "Most Wanted" fugitives listed at the website for the television show *America's Most Wanted.* The names and mug shots are hyperlinks to pages containing more detailed profiles on each wanted individual.

identities of the individuals depicted. The screen interface creates a "navigable space" for the database user, and the software enables automated search capabilities, making it easy to peruse the sites, view the images, read about dangerous identities, and even report back to authorities with more information. Web browsers are themselves software interfaces that allow users to access databases that store the contents of the World Wide Web. Using these browsers, we need little or no knowledge of Internet protocols or other technical standards that make the information retrieval methods work to provide us with search results. In fact, web browsers and search engines are explicitly designed to shield users from the complexity of Internet architecture and information retrieval. Most well-designed software interfaces function in this way, as "user-friendly" tools that allow ease of access to and manipulation of data.

But there are drawbacks to the user-friendliness of the software interface. Most significant, the software interface can become a technology of reification, alienating users from the sources of the information they access and the processes involved in information retrieval, especially the algorithmic techniques that determine the relevancy and ranking of information. As a result, the ease of use of the software interface can invest selective search results with a misplaced concreteness. Users become increasingly disconnected from where the data comes from and the decisions that govern what data is available. As Helen Nissenbaum and Lucas Introna have shown, the seemingly neutral process of information retrieval conceals the politics of search engine design, which tends to overvalue certain sources of information, especially those that are already prominent, while rendering other, less visible sources more obscure and difficult to find.[46] Search engine design decisions can have a powerful impact on search results, defining the parameters of relevancy and the range and rank order of information to which users gain access.

In creating "user-friendly" forms of information retrieval, the software interface and database form introduce new layers of mediation between classification systems and end users of those systems. Like physical filing systems, a database embodies and helps to produce a classification system for the information it contains. And while the database might appear to provide an empirical and technically neutral means of classifying and organizing information, in reality every system of classification is itself a process of social construction that involves particular choices and ways of defining and differentiating classes of objects that do not exist *as such* prior to the development of that system. Classification is a technology visibility, a technology of knowledge and truth production, rather than itself an embodiment of the material reality that it defines and differentiates. Systems of classification "form a juncture of social organization, moral order, and layers of technical integration," as Bowker and Star have argued; they are "artifacts embodying moral and aesthetic choices that in turn craft people's identities, aspirations, and dignity."[47] Classification systems designed into physical filing systems impose an order on their contents in the archiving process. Computerization in turn facilitates the *standardization* of classification systems and their distribution across contexts, multiplying their uses and effects. And as systems become more standardized, the decisions that have gone into devising classification systems recede into the deep structure of technical forms. "When a seemingly neutral data collection mechanism is substituted for ethical conflict about the contents of the forms," write Bowker and Star, "the moral debate is partially erased."[48]

We can begin to see this process of erasure in the way that the *AMW* and FBI websites portray the dossiers of the individuals designated as "most wanted." At these easy-to-navigate websites, the information is presented with authority and simplicity; there appears to be no ambiguity or uncertainty about the identities of the individuals depicted. Clicking on the photograph of a man named Ibrahim Salih Mohammed Al-Yacoub at the FBI website, for example, leads to a page containing the following information in bold, capital letters:

> CONSPIRACY TO KILL U.S. NATIONALS; CONSPIRACY TO MURDER U.S. EMPLOYEES; CONSPIRACY TO USE WEAPONS OF MASS DESTRUCTION AGAINST U.S. NATIONALS; CONSPIRACY TO DESTROY PROPERTY OF THE U.S.; CONSPIRACY TO ATTACK NATIONAL DEFENSE UTILITIES; BOMBING RESULTING IN DEATH; USE OF WEAPONS OF MASS DESTRUCTION AGAINST U.S. NATIONALS; MURDER WHILE USING DESTRUCTIVE DEVICE DURING A CRIME OF VIOLENCE; MURDER OF FEDERAL EMPLOYEES; ATTEMPTED MURDER OF FEDERAL EMPLOYEES.[49]

Just below this information about Al-Yacoub are two black-and-white mug shots, followed by demographic data, a note indicating that he was indicted for the Khobar Towers military barracks bombing, another notifying readers of a five-million-dollar reward, and finally, links to lists of FBI offices and U.S. embassies and consulates. The text presents the identity of this individual in no uncertain terms, and visitors to the site need not question the source or factual nature of the information presented. Al-Yacoub, as described and pictured, is clearly a "terrorist."

Like the cultural representations of the watchlist database found in popular crime and spy dramas, this particular image of the watchlist database does important work for what William Bogard calls "the imaginary of surveillant control"—creating the powerful image (if not the practical reality) of surveillance system capacity.[50] Although visitors to this website can access specific information about the identities of wanted individuals, this simulated version of the watchlist database performs more of a symbolic than a practical role in the identification of terrorists. The site offers up a set of images of the "faces of terror" for the national imagination, and aims to present an authoritative impression of the intelligence gathering efforts of the state. In addition, while neither the *AMW* nor the FBI website makes use of or even mentions facial recognition systems, both sites do important work in defining a compelling social need for facial recognition technology and speak to its conditions of possibility. They underscore the central role of the watchlist database in the design of terrorist identification systems, and the way that the empirical claims to truth about the database depend to a significant extent on technical design strategies that conceal the underlying complexity, ambiguity, and socially constructed nature of classification systems.

**CONSPIRACY TO KILL U.S. NATIONALS; CONSPIRACY TO MURDER U.S. EMPLOYEES; CONSPIRACY TO USE WEAPONS OF MASS DESTRUCTION AGAINST U.S. NATIONALS; CONSPIRACY TO DESTROY PROPERTY OF THE U.S.; CONSPIRACY TO ATTACK NATIONAL DEFENSE UTILITIES; BOMBING RESULTING IN DEATH; USE OF WEAPONS OF MASS DESTRUCTION AGAINST U.S. NATIONALS; MURDER WHILE USING DESTRUCTIVE DEVICE DURING A CRIME OF VIOLENCE; MURDER OF FEDERAL EMPLOYEES; ATTEMPTED MURDER OF FEDERAL EMPLOYEES**

**IBRAHIM SALIH MOHAMMED AL-YACOUB**

**DESCRIPTION**

| | | | |
|---|---|---|---|
| **Date of Birth Used:** | October 16, 1966 | **Hair:** | Black |
| **Place of Birth:** | Tarut, Saudi Arabia | **Eyes:** | Brown |
| **Height:** | 5'4" | **Sex:** | Male |
| **Weight:** | 150 pounds | **Complexion:** | Olive |
| **Build:** | Unknown | **Citizenship:** | Saudi Arabian |
| **Language:** | Arabic | | |

Profile page for Ibrahim Salih Mohammed Al-Yacoub, one of the FBI's "Most Wanted Terrorists." http://www.fbi.gov/wanted/terrorists/teralyacoub.htm.

### *Terrorist Watchlist Database Consolidation*

Although the *AMW* and FBI websites perform more of a symbolic than a practical function in the construction of criminal and terrorist identification systems, the records they display are in fact drawn from actual criminal and terrorist watchlists compiled by state agencies to be used for identifying and apprehending targeted individuals. A closer look at the politics of watchlist database building sheds further light on the classifying logic that informs the application of facial recognition technology for terrorist identification. Programming computers to "see" the faces of "terrorists" would require building a database of facial images that would define the parameters of faces a facial recognition system would identify. But building the terrorist watchlist database has turned out to be a challenging prospect, heavily inflected with the practical politics involved in devising classifications and standards, not least those that aim to assign categories of human identity and standardize those categories across populuations.[51] "Whatever appears as universal or indeed standard," write Bowker and Star, is in reality "the result of negotiations, organizational processes, and conflict."[52] The effort to build a terrorist watchlist database provides a special case in point.

The notion that better use of watchlists may have disrupted the activities of the 9/11 hijackers was a common theme in post-9/11 policy discussions. The 9/11 Commission, for example, was dismayed to find that before the attacks, the U.S. intelligence community did not view building and maintaining watchlists as vital to intelligence work. After 9/11, security agencies built out watchlists with a vengeance, investing more labor, more money, and more machinery into the watchlisting effort. Significantly, more people were placed on watchlists, and the terrorist watchlist became an expanding archive of problem identities. Or more precisely, as several federal audits determined, it was an expanding *set* of archives—separate databases dispersed in different locations, compiled and used by different agencies and actors in ad hoc and inefficient ways. Another central problem underlying the intelligence failures of 9/11, according to government reports and policy discussions, was the lack of consistent information sharing among intelligence agencies. The consolidation of the watchlists among various agencies within the Departments of State, Treasury, Transportation, Justice, and Defense became a pressing political priority.

As the Bush White House began to impose a consolidation program on individual security agencies, the challenge and complexity of database consolidation soon became evident. Two years after 9/11, a General Accounting

DEPARTMENTS, AGENCIES, AND THEIR WATCH LISTS

*(Twelve separate watchlists maintained by five separate U.S. federal departments, to be consolidated by the Terrorist Screening Center.)*

| *Department* | *Agency/Department subcomponent* | *Watch list* |
| --- | --- | --- |
| State | Bureau of Consular Affairs | Consular Lookout and Support |
| | Bureau of Intelligence and Research | TIPOFF |
| Treasury | Customs | Interagency Border Inspection[a] |
| Transportation | TSA | No-Fly |
| | | Selectee |
| Justice | INS | National Automated Immigration Lookout |
| | | Automated Biometric (fingerprint) Identification System[b] |
| | U.S. Marshals Service | Warrant Information |
| | FBI | Violent Gang and Terrorist Organization File[c] |
| | | Integrated Automated Fingerprint Identification |
| | U.S. National Central Bureau for Interpol[d] | Interpol Terrorism Watch List |
| Defense | Air Force (Office of Special Investigations) | Top Ten Fugitive |

Source: GAO.

a. Interagency Border Inspection operates as a part of Customs' Treasury Enforcement Communications System, commonly referred to as TECS.

b. INS is in the process of integrating this system with the FBI's Integrated Automated Fingerprint Identification System.

c. This list is part of the FBI's National Crime Information Center.

d. Interpol (International Police Organization) is an intergovernmental organization made up of 181 member countries for the purpose of ensuring cooperation among the world's law enforcement entities. It is headquartered in Lyon, France. The U.S. National Central Bureau for Interpol, within the Justice Department, serves as the U.S. member of Interpol and facilitates dissemination of Interpol watch list information to federal, state, and local agencies.

Office report found that nine federal agencies were still maintaining twelve separate watchlists, with only sporadic and haphazard information sharing among them. These databases contained a wide variety of data, with no standard or unified set of criteria for who was in the databases, why they were included, or what information about them was retained. Information sharing was hindered by a host of other factors, including not only the "cultural differences" among disparate agencies but also the problem of database incompatibility. The databases of the various agencies were designed for different purposes, by different people, using different software and coding systems, and so were not readily amenable to system integration.

In response to these and other problems, the White House established a new FBI organization called the Terrorist Screening Center (TSC) in September 2003, the latest in a series of consolidation initiatives.[53] The TSC's main purpose would be to develop and maintain a consolidated database, one that had full biometric capabilities and real-time connectivity to all supporting databases. The center would be responsible for making appropriate information accessible around the clock to state and local law enforcement agencies, select private-sector entities, and select foreign governments. By January 2004 the new head of the center, Donna Bucella, testified before the 9/11 Commission that her agency was up and running. Just four months after it was established, the TSC was operating a centralized terrorist database and a twenty-four-hour call center for "encounter identification assistance."[54]

But a Justice Department audit report released over a year later painted a different picture, identifying a litany of problems with the database consolidation effort.[55] The report indicated that the centralized database had a significant number of duplicate records, containing inconsistent and even contradictory information, and that the size of the database differed radically from one phase of development to another. There was a lack of connectivity between the central database and the participating agencies, primarily because many of the agency systems did not support automated data sharing—each of their computer systems would need to be upgraded, which could take years. There were also inconsistencies in the coding of records: for example, over thirty thousand records had "handling codes" that did not correspond to something called their "INA codes." Handling codes indicated what protocols agents should follow in the event of a positive identification with a suspect; INA codes indicated how the individual was associated with international terrorism. Some of the records had handling codes that were too lenient relative to their INA codes, others were too severe, and some of the records had no handling codes at all. The coding scheme in general was not always appropriate to all the records in the database. For example, there was nothing corresponding to an "INA code" for domestic terrorists (who *were* in fact included in the database), so domestic terrorists were automatically assigned international terrorism codes. Significantly, the consolidated database could store biometric data (such as a photograph), but it did not have the capacity to search based on biometrics, so screening continued to rely exclusively on name-based searches. Some known terrorist suspects were missing from the database, and many records were missing important information.[56] The report also noted that some of the records in the new consolidated database were missing the code that designated their originating agency.[57]

As the lengthy list of problems suggested, the watchlist database consolidation effort was a complex, labor-intensive, and highly imperfect process. The reality of watchlist consolidation ran counter both to prevailing popular conceptions of the watchlist database (like those found in popular crime and espionage dramas, and at the *AMW* and FBI websites), and to the seemingly straightforward industry proposals for large-scale, networked terrorist identification systems. While there was certainly a general awareness of the lack of information sharing among U.S. federal agencies, the extent and complexity of the problem of database building was shrouded in a reified notion of the database as an empirical, truth-telling technology. While the database is of course a powerful technology of truth, its truth-telling capacities depend to a significant extent on the way a database is designed, the assumptions behind the classification system that it embodies, and the specific purposes for and conditions under which it is populated. This is not to say that databases cannot yield unanticipated answers, take new forms, or be applied to unforeseen uses. But database building and consolidation are especially challenging processes, fraught with error and ambiguity. No matter how well executed, the database consolidation process itself would have important consequences for the form and content of the resulting classification system. Rather than merely organizing information and offering a transparent reflection of reality, the database form in fact helps constitute the objects it purportedly represents. And consolidating disparate databases involves standardizing their form and contents, thereby transforming those contents in the process. As the head of the Terrorist Screening Center said in front of 9/11 Commission, the individual databases of different intelligence agencies "were created to support the mission of the individual agencies, [they] are in many instances, their case management systems, not terrorist watch lists."[58] In the transfer of data from uniquely devised "case management systems" to the consolidated watchlist, new terrorists were born. What was a "case" particular to an agency investigation became a more generalized "terrorist identity."

In short, one of the most problematic effects of terrorist watchlist expansion and consolidation was a broadened definition of "terrorist" to include a diverse array of problem identities. Like the archive before it, the consolidated watchlist database liberated individual records from the context of their original use, offering them up for new meanings and new uses in the "war on terror." At the same time, the database operated according to an empiricist model of truth, so that this transformation of meaning and use was construed as a technical matter of engineering a more efficient and logical information architecture. The twice-removed status of individual records in a *consolidated* watchlist left the new uses of the archived identities that much further from the context of their original designation as a terrorist or other problem identity. Watchlist database consolidation rendered the "terrorist" designation in technical form, rationalizing and instrumentalizing it as a matter of technical procedure. And the deployment of an ostensibly objective definition of the terrorist identity occurred under the veil of empiricism and technical neutrality that the image of the database afforded.

## *Seeing the Faces of Terrorists*

The difficulties with the terrorist watchlist database consolidation effort underscore the persistent problems involved in building stable identification systems, pushing standardized identity categories back out into the world in order to differentiate populations according to categories of privilege, access, and risk. In spite of every effort at stabilization, argues Jane Caplan, identification systems are fundamentally unstable, precisely because their purpose is to stabilize the inherently unruly concept of identity.[59] Photography was originally applied to identification systems in an attempt at stabilization and modernization, to help visually bind identities to faces and bodies in order to control for the subjective, interpretative, imperfect, and highly variable ways that human beings looked at and identified one another's faces. Automated facial recognition is designed to further assist, and even replace in some cases, the subjective human practice of facial identification, again with the aim of standardizing identification systems and stabilizing human identity. But as the inescapable archival basis of this process suggests most acutely, the instability of identity cannot be resolved through additional layers of technical integration. "Even in its most controlling and technologized forms," identification is "based on a concept that is itself difficult to stabilize and control"—the concept of *identity*.[60]

Despite the seemingly unambiguous ways in which post-9/11 security discourse wielded the trope of the "terrorist" or the "face of terror," there is perhaps no category of identity as unruly and unstable as that of the "terrorist," a fact that comes into sharp relief in the case of the terrorist watchlist consolidation effort. Devising a consolidated watchlist database for terrorist identification was a biopolitical project, one that inescapably bound the security of the population to the state's sovereign claim to the right to kill. Killing here meant not just to literally murder people but also to increase the risk of death for some people, inflicting "political death, expulsion, rejection, and so on."[61] Although automated facial recognition has not quite materialized as a functioning technology for terrorist identification, it nevertheless promises to facilitate precisely these biopolitical aims, working with expanding databases of mug shot images. But regardless of whether individuals designated as terrorists can be identified using automated forms of facial recognition, the *virtual* face of terror serves a key biopolitical function, brandished as a weapon to justify state racism and define the war on terror as a virtuous one.

The drive to stabilize identity by automating the perceptual labor of identification connects innovations in identification systems with debates about the nature of visual perception and its relationship to photographic technologies. Scholars of visual culture have debated at some length the implications of new visual media technologies for our understandings and experiences of sight and vision. Does the automation of vision open up our experience of the visual world to new possibilities, or is machine vision having repressive and even blinding effects on how we see the world? Paul Virilio has perhaps the bleakest diagnosis.[62] In the drive to automate visual perception, he sees inherent imperialistic and militaristic impulses. The technologies of war and cinema have been developed in concert, enabling both the visual augmentation of military logistics as well as the militarization of societies and their mobilization for war. If machines can "see" in Virilio's analysis, they have a frightening tendency do so with an eye for bloodshed. Kevin Robins similarly sees inhuman and inhumane propensities in new visual media technologies. For him, they embody a desire to disavow the reality of the world's events and place distance between people and the worlds they inhabit. "Technologically mediated vision," writes Robins, "developed as the decisively modern way to put distance around ourselves, to withdraw and insulate from the frightening immediacy of the world of contact."[63] Automation goes hand in hand with the rationalization of vision, according to this view, and the development of prosthetic visual devices separates visual perception from the natural capacities of the human body. The "relentless abstraction of the visual" through computerized imaging technologies has involved "relocating vision to a plane severed from a human observer," robbing humans of their privileged status as perceiving subjects.[64]

Others take a different view of technologically enhanced vision. John Johnston has argued that critics like Virilio wrongly assume a stark opposition between the human and the technical, and this assumption imposes limits on reflection about vision and visual culture.[65] Johnston does not believe that human vision can be opposed to machine vision, instead using the Deleuzian concept of "machinic vision" to theorize "an environment of interacting machines and human-machine systems," as well as "a field of decoded perceptions that . . . assume their full intelligibility only in relation to [machines]."[66] In other words, new human-machine assemblages create new possibilities for the subjective experience of sight, for what vision means and how it works, and for how humans and machines collectively see the world. Donna Haraway similarly argues that humans are already inescapably embedded with visual media technologies. It is in the intricacies of these embedded relationships between humans and technologies, she argues, "that we will find metaphors and means for understanding and intervening in the patterns of objectification in the world, that is, the patterns of reality for which we must be accountable."[67]

Johnston and Haraway are right to argue that human visual perception is inescapably integrated with technical forms, that visual forms of perception are constituted not by human bodies alone but through assemblages of human vision, visual devices, and techniques of observation. But whether humans will ever use the changing vantage points afforded by new forms of "machinic vision" to see the world differently remains to be seen. In the process of designing new technologies of vision, our subjective forms of visual perception are themselves being reinvented. What seems cause for concern about this process is how rarely (if ever) it leads to radically new ways of seeing, instead reinscribing, in more effective ways, existing power relationships and dominant modes of vision. Any celebration of the radical possibilities opened up by new human-machine visual assemblages must be tempered by a recognition that a certain set of institutional actors and their constituencies are the primary agents and beneficiaries of emergent forms of "machinic vision." It may in fact be misguided to hold out a nostalgic attachment to the natural capacities of human perception unadulterated by mediated forms. But it is likewise mistaken to turn a blind eye to what are clearly the dominant forms of human-machine perception, forms that incorporate and authorize very particular ways of seeing the bodies and faces of others.

In their development and practical applications, automated facial recognition systems make use of socially constructed classification systems for defining and standardizing identity categories. Formalizing these classification systems, amassing databases of facial images, and implementing new identification technologies are institutional projects being undertaken mostly by state agencies, law enforcement actors, and business entities. But there are also ways in which these projects and their associated practices are taking shape at the level of the individual, imported into the realm of everyday practice for a certain class of security-conscious, tech-savvy citizens. Individual users are participating in the project of facial image database building, taking photographs of faces, storing them in databases on their computers, and migrating them online. They are also being invited to participate in the development of facial recognition technology by experimenting with software prototypes designed for individualized use. The next chapter examines how the securitization of identity is imported into the self-fashioning activities of tech-savvy individuals through the introduction of consumer applications of facial recognition technology.

# 4

# Inventing the Security-Conscious, Tech-Savvy Citizen

Welcome to MyFaceID, your personal facial recognition service!
—http://www.myfaceid.com/, June 25, 2008

In early 2008, Lenovo, the company that acquired IBM's PC division, released a new line of laptops equipped with VeriFace facial recognition software. Instead of using a password to log in to their computers, users of these laptops have their pictures taken and verified against previously enrolled facial images. The software supports multiple user accounts and keeps a log of photos of everyone who attempts to access the computer. Also in 2008, a German software company called Betaface introduced an online facial recognition search engine, called MyFaceID, that allows users to upload photos of people's faces and match them against other faces in the MyFaceID database. Along with its web release, Betaface launched a platform application for the social network site Facebook, which, according to product descriptions, subscribers could use to automatically process their photos, find faces, "tag" them with the names of the people depicted, and search for facial resemblances.[1]

Previous chapters have focused on applications of facial recognition technology designed for institutional users—how the needs of businesses enterprises, police departments, and state agencies to secure networks and enhance their surveillance and identification systems have shaped the effort to program computers to "see" the face. As we saw in chapter 1, the deployment of biometric systems has been a protracted process taking shape across a range of sectors, from the banking, finance, and credit card industries to state DMV offices, prisons, welfare agencies, and border and immigration control systems. Despite the best efforts of proponents, the desirability of automated facial recognition and other types of biometrics has been far from self-evident to everyone. One of the key obstacles to more widespread implementation of biometric technologies has been what the biometrics industry refers to as "user acceptance." The institutionalization of biometric identification requires the habituation, co-optation, and cooperation of end users.

This chapter seeks to examine how social uses of facial recognition and other biometrics are being defined at the level of individual users, and how the securitization of identity is incorporated at the level of individual practice. While biometric identification systems are being envisioned and designed mainly for large-scale, database-driven, institutional uses, developers have experimented with consumer applications—hardware and software systems designed and marketed for individual users. These consumer applications help to enlist the cooperation of individual users in the process of biometric system development and institutionalization. Although consumer applications are not being developed for the express purpose of enlisting user acceptance of biometrics, they nevertheless help to put the technologies into practice at the level of the everyday, self-fashioning activities of "tech-savvy citizens"—people for whom technical competency and active participation in their own self-government are two mutually constitutive responsibilities. Responsible, security-conscious citizens are expected to not only own and make use of a wide range of digital technologies in their everyday practices, but also to secure, manage, and optimize their devices and personal data. Consumer applications of biometrics are designed with the new responsibilities and competencies of the security-conscious, tech-savvy citizen in mind.

To explore how consumer applications of biometrics link up to the needs, desires, and demands of tech-savvy citizenship, I examine two types of consumer applications. First, I look at personal security applications—shrink-wrapped versions of facial recognition and other types of biometrics, designed and marketed as tools for helping individuals secure their own property, devices, and data from unauthorized access or theft. Here automated facial recognition is cast as a technology that can help individuals protect their privacy and property. Consumer biometrics are included among the range of requisite techniques for securing oneself from theft, not only of one's electronic devices and other belongings, but also theft of one's identity itself. These personal security applications embody the logic of "individualized security" associated with neoliberal political rationality, posited as a means of enabling individuals to take responsibility for their own security.[2]

The second type of consumer application I discuss in this chapter is facial recognition software designed for the management of personal photo collections and online photo sharing. By 2008, facial recognition software prototypes were accessible on social networking sites, designed to enable users to manage their personal photo collections and search for faces among other users' profiles. As consumers have adopted digital cameras and computer storage, the quantity of photos that they take has increased exponentially, making the organization of personal photo archives an especially challenging task. Moreover, online photo sharing sites like Flickr, Riya, and Picasa, as well as social network sites like MySpace and Facebook, have enabled these vastly expanding photo collections to migrate online, giving them a decidedly more public presence. Photo management applications of facial recognition technology have been envisioned as a way of helping users identify faces in images, organize their photos by faces, and establish connections based on automated visual comparisons. Google's 2006 purchase of Neven Vision, a tech firm specializing in automated object and facial recognition, was one indication that the major brokers of the online economy were beginning to take seriously the application of facial recognition technology for photo archive management. As we will see, Internet users are not only invited to adopt online prototypes of facial recognition software, but also to experiment with these beta software applications and offer feedback to developers, thereby participating in the development and evaluation of the technology. This model of participatory, "interactive" involvement in the design of facial recognition software enlists end users to contribute their free labor and technical skills to the process of developing more effective forms of image retrieval, using social network and photo sharing sites as test beds for defining new social uses of facial recognition technology and optimizing its functionality.[3]

The claim of "empowerment" pervades discourse about digital technologies in general, and consumer applications of biometrics are no exception. The suggestion is that we can "empower" ourselves by optimizing our personal technologies, trying out and adopting the latest new devices, software, and services in order to better manage our lives. We are encouraged to experiment with and make optimal use of new technologies, whether to express ourselves, increase our productivity, make more money, secure ourselves from threats, or otherwise better ourselves and make our lives more self-sufficient and complete. But these ceaseless efforts to convince individuals of the need to adopt new technologies in pursuit of empowerment and completeness deserve critical attention. Biometric systems are being designed primarily to invest institutional users with the enhanced capacity to control networks and information, to increase the scale and effectiveness of surveillance systems, and to secure forms of differential access to the resources of value in late capitalist societies. Despite the language of empowerment, consumer applications of biometrics support rather than challenge this aim.

## *Tech-Savvy Citizenship*

Programming computers to recognize human faces and facial expressions is a project that cannot help but have ties to contemporary notions of subjectivity and citizenship. Almost every new technology that we integrate into our lives signals a change, however slight, in our sense of what we can do, what we should do, and who we are as modern citizen-subjects. The philosopher Langdon Winner notes that "to invent a new technology requires that (in some way or another) society also invents the kinds of people who will use it. . . . Older practices, relationships, and ways of defining people's identities fall by the wayside; new practices, relationships, and identities take root."[4] The situation is of course more complicated than this, since we also design new technologies in order to help us do particular things and be particular kinds of people, and since these changes play out in complicated ways, with plenty of unintended consequences. But it nonetheless holds true that human beings are in some ways invented and reinvented along with the invention of new technologies. We had to learn to become people who drive cars, for example, and cars have contributed to the complete transformation in our way of life and who we take ourselves to be.

Facial recognition and other biometric systems are being made to function as *identification* technologies in a double sense, both as bureaucratic forms of official identity verification and as mechanisms that tech-savvy citizens can make use of to define, maintain, project, and protect their identities. In their search for markets for automated facial recognition, developers are experimenting with personal security and personal photo management applications, articulating the technology to the needs and responsibilities of an ideal-typical security-conscious, tech-savvy citizen. These individualized uses of the technology are particular forms or aspects of what Foucault calls "technologies of the self," techniques that "permit individuals to effect by their own means, or with the help of others, a certain number of operations on their own bodies and souls, thoughts, conduct, and way of being, so as to transform themselves in order to attain a certain state of happiness, purity, wisdom, perfection, or immortality."[5] Foucault did not use the term "technology" in reference to electronic devices or software, but there is a sense in which the varied uses of computers, cell phones, MP3 players, e-mail, profiles on social network sites, and other hardware and software technologies are part of the "technologies of the self" for a certain class of privileged, tech-savvy individuals. Social network sites enable users to engage in practices of self-monitoring and self-actualization, while also providing a means of keeping tabs on one another. More than electronic gadgets, our new personalized devices are part of the repertoire of mechanisms that we use to communicate and reflexively constitute ourselves, and especially to make our lives more productive, efficient, and entrepreneurial. A command of these technologies, and continuous experimentation with upgrades and new innovations, likewise has become part of the expectations of modern citizenship, and the pressure to adopt new devices and acquire new skills—as well as the pleasure derived from doing so—has become a significant force of social mobilization. As Andrew Barry has argued, "Technology itself, in the form of networked devices, is thought to provide a significant part of the solution to the problem of forming the kind of person who can exist, manage, compete, experiment, discover, invent and make choices in a technological society."[6]

Consumer applications of facial recognition and other biometrics are precisely such networked devices, envisioned as part of the solution to the problem of responsible citizenship in a technological society. They are examples of the kinds of technologies that tech-savvy citizens are expected to experiment with and adopt in order to optimize their technical devices—to make them work more effectively and to make the most productive use of them. The integration of facial recognition and other biometrics with personal-use technologies provides a way of further optimizing their potential and enhancing their functionality as the prosthetic devices of identity. Experimenting with new technologies doubles as a form of experimenting with the self, and making optimal use of technologies becomes a means of optimizing our lives, testing and adopting a never-ending stream of innovations as a way of inventing and reinventing ourselves as a capable, competent, productive citizen-subjects. Of course, this experimentation is not unbounded; while we might be encouraged to experiment with biometrics, these technologies are ultimately designed to *fix* our identities, to bind our identities to our bodies and make us identifiable across space and time. The flexible, experimental subject is *always identifiable*, and managing her legitimate, legal identity is absolutely essential to her ability to function as a productive citizen living in a modern society. The flexible, identifiable subject doubles as an ideal-type of citizen, one who does her part in the broader effort to test and evaluate nascent technologies, offer feedback to developers, secure her devices and data, and use new devices and services to monitor her own conduct and the conduct of others.

## *Personal Security*

Much of the discourse about biometrics positions the technologies as ideally suited for the protection of individuals' "personal security," a concept and set of practices that gained new currency along with the rise of neoliberalism. During the 1980s and 1990s, postwar social welfare programs designed to provide some measure of economic security for individuals and families became subject to criticism, and were replaced or modified to a significant extent by entrepreneurial models of self-motivation and self-reliance. The business of dismantling social welfare programs was accompanied by a pronounced "individualization of security"—a new level of emphasis on the responsibility of individuals and families for their own security and risk management, along with the propagation of an image of state-centered forms of social welfare and collective forms of economic security as socially degenerative.[7] Although not a radically new idea, individualized security began to take on a heightened sense of importance under the policies of Reagan and Thatcher in the 1980s. Nikolas Rose and others have focused on private insurance as a key site for understanding the new politics of individualized security and risk, but other industries and technologies have likewise emerged in response to this governing logic. Personal security applications of biometrics embody and enact the logic of individualized security, working from the assumption that individuals should take responsibility for their own security, including the security of their bodies, property, devices, and personal data.

As a general problem, the very idea of personal or individual security has deep ties to both one's sense of self and to one's officially established identity. ID documents are themselves "personal security" technologies, and the very idea of "personal security" is inescapably tied to our officially established identities—the identities that appear on our ID documents. It has for some time been the duty of individuals to secure the appropriate identification documents for authorizing their transactions with institutions and otherwise establishing their status as legal persons. These documents authorize our transactions not only with the state, but with a wide range of other institutions, especially financial institutions. We are obliged to carry our documents with us, to make sure they remain valid (unexpired with updated address and other information) and in readable condition, and to protect them from duplication, alteration, or unauthorized use. While the standardization of identification documents and the integration of new technologies into both the documents themselves and their processing are beyond the scope of individual responsibility, it is nevertheless each citizen's duty to make sure that she has the most up-to-date, standardized version of identity authentication. Taking the time to visit the appropriate agencies and manage one's official documentation is part of the responsibilities of citizenship and absolutely essential to personal self-management.

As noted in chapter 1, industry rhetoric promoting the biometric driver's license often claims that the new high-tech documents will make life easier, safer, and more secure for law-abiding citizens. Industry and policy discussions about the integration of biometrics into standardized identification documents present this transition to biometric identification as beneficial to documented citizens, a way of enabling them to continue to move about freely and otherwise engage in normal daily activities in the face of pervasive risks to their well-being. Advocates of biometric identification documents often suggest that the new technologies make the documents more secure by protecting the bearers' privacy and personal data. Decisions about the design features of identification documents occur at the institutional level, especially to satisfy demands for more secure forms of identification such as those coming from the banking and credit card industries, but the justification for biometric integration is often framed as a matter of citizen or consumer protection and benefit. And security-conscious, tech-savvy individuals in fact have a vested interest in the security of their financial identities (although whether in fact biometric technologies are effective in protecting financial identities is by no means certain).

One particular selling point for biometric driver's licenses and other personal security technologies has come from heightened press and policy attention given to the issue of identity theft. While proponents of biometrics maintain that enterprise-level applications are necessary to provide protection to individuals, there is also special emphasis on the responsibility of individuals themselves to adopt biometric-integrated devices on their own as means of protecting their devices and identities. This emphasis on individual choices and responsibilities is prevalent in identity theft discourse, which often stresses individualized rather than systemic forms of prevention.[8] According to Simon Cole and Henry Pontell, the problem of identity theft has been individualized and privatized, with individuals expected to take measures to protect themselves, for example, by buying "credit protection" services, paper shredders, and biometric-integrated devices.[9] But protecting one's identity can be far more difficult than protecting one's physical property, as Cole and Pontell explain, since the entity to be protected takes the less tangible form of information, and that information is dispersed in remote databases that lay beyond any individual's control. The measures individuals are supposed to take to protect themselves from identity theft are extremely limited in terms of how much control they actually give individuals over the security of their data and identities.

While the personal measures that people are encouraged to take to protect their identities may have limited effects on actual data security, the expectation that individuals take responsibility for protecting their identities does have the effect of creating markets and generating revenues for industries in the business of personalized identity theft protection, including consumer applications of biometrics. But consumer biometrics are not always sold as stand-alone products. The integration of biometrics with consumer devices like laptops and cell phones takes place at the level of production and design. In other words, consumers are not called on to integrate biometric systems themselves; instead, they are encouraged to choose to purchase digital devices that have integrated biometric capabilities. The company AuthenTec, for example, bills itself as "the world's leading provider of fingerprint authentication sensors," and identifies its customers not as individual consumers or end users, but manufacturers of consumer electronics, like Fujitsu, HP, LG, Samsung, and Toshiba.[10] But while manufacturers represent the company's primary market, AuthenTec devotes considerable attention to their indirect market of end users. Their website displays a series of images of people using devices, with accompanying tag lines referring to "convenient security . . . at work . . . at home," and " . . . for people on the go." Additional copy refers to the "added security and convenience that fingerprint biometrics bring to *everyday life*."[11] AuthenTec also conducts annual surveys of consumer attitudes toward biometrics and posts the results on the main page of their site.[12] Recent survey results (2008) predictably show a majority of respondents favorably disposed to using biometric devices with their personal computers and cell phones, noting that 77 percent of respondents "are ready to begin using fingerprint sensors as part of their day-to-day activities," and that 68 percent "trust fingerprint biometrics more than traditional passwords."[13]

Like all industry-sponsored studies, these survey findings reflect the interests of the industry and so should not be read on their own terms.[14] Despite what the industry-sponsored research suggests, consumers are not amassing in large numbers to make vocal demands for biometric-secure devices. Instead, the biometrics industry is attempting to *create* a market for biometric-integrated computers and cell phones, in this case by persuading consumer electronics manufacturers that a growing number of consumers are recognizing the need for biometric security. For their part, the developers of PCs and cell phones have a vested interest in integrating biometrics as part of their general effort to perpetuate both technological and stylistic obsolescence. The consumer electronics industries continuously innovate and otherwise modify technologies to make existing computer and wireless devices either seem outmoded or actually fail to meet the system requirements for new applications. They also expend considerable energy convincing users that their existing equipment has a limited time frame of functionality, and that every new device they purchase will soon be inadequate to meet their ever-expanding needs and levels of technological sophistication. No doubt tomorrow's obligatory fingerprint readers will quickly be replaced by even more robust and secure biometric sensors. In fact, a seemingly inadequate biometric security device could even serve as reason for replacing a functioning computer or cell phone, especially since even the tech-savviest of users can hardly be expected to actually do the work of integrating biometrics with their personal devices themselves. Responsible, security-conscious consumers will be compelled to scrap their functioning devices for shiny new and more secure versions of the same.

Biometrics are being positioned as part of the essential requirements for other aspects of self-management and self-protection as well, including home security. As James Hay has shown, the modern domestic space represents a special site of social regulation and self-government, a primary setting where individuals put recommended "designs for living" into practice.[15] Home security systems have become essential components of such designs for living, with signage for ADT and other private security firms littering the suburban lawn-scape. Biometric technologies are being marketed along with burglar alarms, video cameras, and DVR machines to upscale and other security-conscious home owners. There are no reliable statistics on the number of home biometric systems in operation, but it is safe to assume that a small but growing number of homes are adopting them. Fingerprint door locks are the most commonly marketed devices (prices at the time of writing ranged from about $200 to $1,600). Although replacing key locks with biometric access-control devices represents a relatively minor change to conventional home security practices, these biometric security systems are innovative to the extent that they aim to bind the security of the home to the bodies of its authorized occupants, providing home owners with an instrument of biopower for the home. More sophisticated home biometric systems provide not only a way of keeping intruders out of the home, but a way of monitoring the comings and goings of the people who live there.

If biometric home security systems make sense in a climate of individualized security, they also make sense insofar as they articulate "home security"

"Armed Response" signage in front of a security fence surrounding a home in an upscale neighborhood in La Jolla, California. Photo by Kelly Gates.

to "homeland security," giving the home an aura of high-tech access control and "civilian preparedness." Reimagining the bomb-shelter versions of home security advocated in the Cold War context, "homeland security" discourse emphasizes the home as "the first line of defense" in the "war on terror," making the security of the home the primary responsibility of its inhabitants, with some help from the advice of "security experts."[16] Although there is nothing new about assumptions that people should protect themselves and their homes from external threats, the rise of an ethic of individualized security has been accompanied by an intensified focus on the home as a site of personal safety and security. The new "smart home" is a *secure* home, fully equipped with an array of safety devices, from smoke detectors and timers for lights to security alarms wired to private security firms and even private "armed response" officers ready to respond to home security breaches. (For a critical mass of "armed response" yard signage, see for example the ocean-view neighborhood in La Jolla, California, that borders the western edge of the University of California, San Diego, campus.)

Biometric home security is not only for home owners. Developers of gated communities and condominium complexes are integrating biometric access control into the design of security systems in order to assure that these privatized spaces remain accessible only to residents and guests, and also to reduce the labor costs of employing security guards and doormen. In these cases, biometric systems are usually part of the total security package contracted out to private security companies. Individuals get prepackaged home security along with the range of other benefits that come with condo living, like having the lawn mowed and the pool cleaned. For consumers of these private developments, biometric home security is part of a broader set of choices that they purchase from a menu of options. The very decision to live in a gated community, fortified condo complex, or secure building becomes a "lifestyle choice" that doubles as a sign of sensible, responsible citizenship. (The implication is that if individuals choose to live in vulnerable, unprotected apartment buildings or freestanding houses in unfortified neighborhoods, they do so at their own risk.)

Biometric security systems are also found in low-income housing developments. These compulsory systems are tied less to the responsibilities of citizenship and self-government, however, than to the flip side of neoliberal control strategies: enforcement of inequality and differential access. Early examples included Marshall Field Gardens and Skyline Towers in Chicago, both of which installed hand geometry biometrics in the 1990s as part of an intensification of security and access control in "distressed housing."[17] Although the actors making decisions to install biometric systems at these sites no doubt have some concern for the safety of residents in mind, another stated purpose is to monitor unauthorized occupancy and adjust rental assistance allocations.[18] The U.S. Department of Housing and Urban Development (HUD), which allocates rental assistance, also distributes grants to fund biometric installations. Descriptions of biometric access control systems at low-income housing developments invoke prison-like security.[19] At Marshall Field Gardens, for example, "single-passage portals" use a system of interlocking doors to allow authorized residents to enter only after their identities have been verified by hand scan, and audit trails created by the biometric system have been used in police investigations.[20] As descriptions of these systems suggest, the coercive participation in biometric security at low-income housing developments has more in common with the biometric registration of welfare recipients, criminal suspects, and prisoners discussed in chapter 1 than with the lifestyle choices associated with more upscale versions of gated communities.

Perhaps more important than creating a market for personal security technologies, the hyper-security-conscious climate engendered by moral panics surrounding things like identity theft, home intrusion, and terrorism also encourages people to embrace *institutional* forms of biometric registration. Undoubtedly, one of the difficulties that many people have had with biometrics—and one of the most significant impediments to what the biometrics industry refers to as "user acceptance"—concerns the way these technologies seem to interface with the body in a more direct way. Biometrics seem to intervene on the sense of control that people must have over their own bodies as a matter of their psychic well-being. The idea that we have control over our own bodies is at the heart of human agency as both a concept and a possibility. Lack of control over the body—a matter with deep historical ties to gender and racial inequality—is linked to the fragmentation of subjectivity (which is the goal of torture techniques, as Naomi Klein has shown).[21] The relationship between self-identity and control over one's body represents one of the many issues that falls under the umbrella issue of "privacy," but this definition is often obscured in favor of other conceptions of privacy, such as the right to control information about oneself, the right to legal protections against search and seizure of property, or vague notions about the right to be "left alone."

While the prospect of widespread deployment of biometric technologies seems to pose a new level of threat to the value of privacy defined in terms of individual autonomy and personal control over the body, the possibility that individuals themselves may choose to adopt biometrics to serve their own security needs would seem to alleviate that threat, at least to some extent. In other words, consumer applications of biometrics promise to invest individuals with a sense of control over these technologies, rather than feeling controlled by them. As consumers gain a measure of technical skill in the use of biometrics, and as they become more commonplace technical forms integrated into their everyday practices, it seems logical that people would view the technologies as less threatening, and in fact come to see them as a necessary part of their own responsible, security-conscious conduct. Moreover, experimental consumer applications of biometrics may help people feel more comfortable not only using biometrics themselves, but also being subjected to institutionalized forms of biometric registration and coming in contact with larger-scale biometric systems on a regular basis. If people are able to experiment with their own personalized forms of biometric security, then the biometric systems deployed by their employers, financial institutions, state agencies, and other institutions seem less like an invasion of privacy and more like obvious and necessary measures being instituted for their own protection.

### *Personal Photo Management*

In addition to "personal security" applications, facial recognition technology has also been incorporated into search engine and photo management software aimed at individual users. In 2008, facial recognition software prototypes began to appear as platform applications on social networking sites like MySpace and Facebook, designed to allow users to upload and perform matches on their own photos, and photo sharing services like Riya and Google's Picasa began to incorporate facial recognition functionality. Soon thereafter, Apple released a version of iPhoto with a "Faces" feature, a facial recognition function that allows users to organize their photos by faces. Also, as part of their promotional campaign for Coke Zero, Coca-Cola Company launched the "Coke Facial Profiler" on Facebook. It allows users to upload their photos to search for similar faces on the web. "If Coke Zero has Coke's taste," asks the interactive advertisement, "is it possible someone out there has your face?"

As these proliferating software prototypes suggest, facial recognition technology is being tailored to address the new set of challenges posed by the availability of cheap digital cameras and increasing computer storage and processing power—one means of managing the avalanche of images flowing into databases and onto the Internet. Integrating facial recognition technology with photo management software promises to rein in this chaotic and expanding field of images, giving individuals access to an "interactive" means of identifying individuals depicted in photos and searching for likenesses among other users' photo collections. Internet users are invited to experiment with these beta software applications for managing their photo collections and making their images more searchable, and in turn offer feedback to developers on how to improve the technology and adapt it to suit their needs.

Two online prototypes, the "MyFaceID" Facebook application and the "MyHeritage Celebrity Look-Alikes" application on MySpace, have similar features but take somewhat different form. Both companies are developing facial recognition technology as part of a business model and range of product offerings that extend beyond their social networking applications. MyHeritage.com is a company based in Israel that offers online family genealogy services, with an interest in adapting the technology to the task of making genealogical connections among subscribers' family photos. Betaface, the German company developing MyFaceID, offers media asset management services, specializing in information retrieval software designed to search for faces, speech, and text in volumes of multimedia content. Both the MyHeritage and MyFaceID applications have their own websites where visitors can experiment with the software in much the same way that they can on Facebook and MySpace. While both applications encourage user experimentation with facial recognition search engines, they differ in the kind of experiments users can perform. As its name indicates, the MyHeritage Celebrity Look-Alikes application offers a way of searching for resemblances among users' uploaded photos and images of famous people. Users can then post the results of their look-alike searches on their MySpace profiles. The MyFaceID application, on the other hand, is promoted primarily as a way of finding likenesses among photos of ordinary people. In other words, it allows users to match their photos against those uploaded to the MyFaceID image database by other users. The description of MyFaceID indicates that the application allows you to "automatically process your photos, find all faces, help you tag them and let you search for similar people."[22]

The emerging social uses of these photo management technologies have genealogical connections to the history of innovations in everyday, popular uses of photography, beginning with the contribution photography made to the democratization of portraiture. As the historian John Tagg has argued, the rise of the photographic portrait was one response to the growing social, economic, and political importance of the middle class in the nineteenth century: "To 'have one's portrait done' was one of the symbolic acts by which individuals from the rising social classes made their ascent visible to themselves and others and classed themselves among those who enjoyed social status."[23] What seemed so wonderful about the new mechanical portraits was not only how fast and cheaply they could be produced relative to drawings or paintings, but also their unprecedented accuracy and authenticity. They provided an automatic, mechanically transcribed, and meticulously detailed likeness of the face, which in turn seemed to provide a more direct connection to the people depicted than a drawing or painting.

With George Eastman's innovation and heavy promotion of the Kodak camera, photography underwent a major transformation. As cheaper "Brownie" cameras were produced and marketed to those of modest incomes, many more people began taking pictures themselves and arranging "the intimate, informal or ill-composed results" in personal picture albums.[24] As Susan Sontag has argued, photography became a central part of the social rituals of family life at the very moment when the nuclear family was beginning to take precedence over a more extended social organization of family members. Photography helped to fill the gap left by this rupture: "Photography came along to memorialize, to restate symbolically, the imperiled continuity and vanishing extendedness of family life."[25] What emerged was the family album, an archive of images that family members took of themselves and one another, compiled together to stand in for real togetherness.

Amateur film photography also developed in tandem with the popular leisure activity of tourism, Sontag notes, helping to dispel some of the anxieties associated with travel by placing a level of mediation between tourists and the alien places in which they found themselves. Taking pictures afforded amateur photographers a sense of control over the settings they captured on film.

But it is not quite accurate to view the rise of popular uses of photography as simply giving more power to the people. What happened, according to Tagg, was a "far-reaching pictorial revolution," a reversal in "the political axis of representation," so that it became less of a privilege to have one's portrait taken than a "burden of a new class of the surveilled."[26] As Sekula has argued, photography functioned not only by extending the "honorific" conventions of portraiture downward to the middle classes, it also functioned repressively, providing a new instrument for social classification. By introducing the panoptic principle into daily life, "photography welded the honorific and repressive functions together."[27]

Like these earlier developments, digital photography and its migration online must be understood in relation to a set of contemporary transformations in everyday, visual-cultural practices, and especially the way these practices weld together institutional and personal uses of new visual media technologies. Technological developments in image capture, accumulation, and sharing are being adapted to the perceived needs and desires of a newly mobilized and meticulously individualized populace. People are encouraged to make optimal use of digital cameras and online photo sharing software to express themselves visually, enrich their online profiles, establish connections, and build and maintain mediated relationships. The direct connection that early photography seemed to establish between the mechanically produced visual likeness and the actual embodied person resonates in the world of online social networking and photo sharing, as users browse one another's files in search of traces of themselves and people they know. The exciting part about online photo browsing is not so much the detail and accuracy of images, however, but the volume and accessibility of other peoples' photos, what people's photos tell us about them, and the possibility of discovering photos of interest in a sea of images. The relationship between image and reality remains an issue, to be sure, but it exists largely as a question of verification rather than as a technical accomplishment. The social connection once afforded by the automatic reproduction of a detailed likeness of the face is overshadowed by the need for an automatic means of finding and identifying people online.[28]

To be sure, the mediated experience of encountering people online has raised new problems of identity verification, and online photosharing plays an integral role in both the construction of the problem and the solutions envisioned to address it. People are not always who they claim to be on the Internet, and the pictures they post are often viewed as the most potentially deceptive aspect of their online identity performance. This is not a trivial issue, since confidence in the reliability of others is part of the basic conditions of trust essential to individuals' ontological security—the practical and largely unconscious ways that we are able to manage our normal, day-to-day existence, more or less bracketing the ever-present possibility of disaster. "Trust in others. . . is at the origin of the experience of a stable external world and a coherent sense of self-identity," writes Giddens.[29] In modern societies, trust has increasingly become a matter of verification, and especially in the online context, verification becomes something that requires technical intervention. Ironically, the social anxieties brought on in part by new technologies call for additional layers of technology to help dispel those anxieties.

Much like its application for more secure identification systems (like the driver's license discussed in chapter 1), automated facial recognition is envisioned as a technical means of establishing the trustworthiness of others in communicative contexts characterized by an absence of embodied, face-to-face interaction. Facial recognition technology uses the iconicity of facial images as a means of establishing their indexicality, their direct relationship to real, embodied individuals. The intensified problem of trust inherent in disembodied interactions is especially salient in online contexts, and facial recognition software is now being introduced as a technology that Internet users can adopt to address this problem. It promises to provide an authoritative, technical means of establishing whether individuals are who they claim to be online. But the potential uses of facial recognition technology for photo management are not limited to helping users verify the identities of individuals online. The technology also promises to help answer more open-ended questions about individuals depicted in photos—not only "is this person who she says she is?" but also "who is this person?" or more broadly, "who are all these people?" Facial recognition search engines might also help locate photos of persons of interest ("where can I find additional photos of this person?") or, as in the MyHeritage Celebrity Look Alikes application, people who merely resemble one another ("who does this person look like?").

In other words, facial recognition software prototypes promise users new ways of interacting with their own photo collections and other collections they encounter online, in this way embodying the logic of new media "interactivity." This model of interactivity shares important affinities with the one pioneered by science museums in the 1990s. According to Andrew Barry, science centers and science museums were among the first institutions to develop programs that promoted the idea of interactive technological experimentation.[30] Interactive science exhibits were designed to encourage museum visitors to engage in a form of "play," using their bodies and senses to gain a better understanding of scientific principles. In this way, these exhibits promised to enliven more conventional science museums and to transform the museum visitor into "an experimental self," appropriately curious about science and interested in engaging in the process of discovery.[31] Investing individuals with an interactive sensibility would encourage them to keep abreast of technical innovations and continuously acquire new technical skills. While individuals would have to invest considerable labor in reeducation and skills acquisition to keep pace with a changing technological environment, the model of an interactive, experimental self framed this process as an exciting and pleasurable undertaking rather than an onerous act of submission to new forms of discipline.[32]

A related promise associated with the logic of new media interactivity is that new media enable greater user participation in the production of media content, challenging the passive forms of consumption associated with conventional broadcast and print media.[33] But the participatory promise of this celebrated form of "interactivity" belies a fundamental reality. As Mark Andrejevic has argued, the agency that users seem to gain in networked communication environments is accompanied by intensified monitoring of their activities, with the ultimate aim of helping marketers optimize commercial messages and target them more effectively to specific audiences. Although new media technologies are often said to "empower" individuals by enabling them to provide input and create and present their own content, interactive media also compel users to willingly submit to being monitored as a condition of use. "The defining irony of the interactive economy," writes Andrejevic, "is that the labor of detailed information gathering and comprehensive monitoring is being offloaded onto consumers in the name of their own empowerment."[34] This dual function of interactive media is reminiscent of the combined honorific and repressive functions of early photographic portraiture: participation in the online economy is at once a privilege and a burden of tech-savvy citizenship.

One way that photo management software prototypes embody the reflexive logic of interactivity is by encouraging the habituation of photo tagging, a practice that plays a fundamental role in automated forms of image retrieval.

For most of its short history, computerized image retrieval has involved text-based techniques. When we search for images online, we enter keywords, and the search algorithms compare those keywords against the metadata associated with archived photographs, rather than using the actual visual information contained within the photographs themselves. Dates and other basic information can be automatically attached to photographs as metadata, but for the most part, any specific information about the content of images must be put there by human beings doing the (unpaid) labor of data entry. Searching for photos of specific people online can be especially difficult and time-consuming, because online databases now contain millions, perhaps billions, of photographed individuals, and only those "tagged" with the names of those individuals are searchable using standard, text-based image retrieval techniques.

The practice of photo tagging, or adding metadata to photos to identify people, is essential to computerized image retrieval, with individual users doing the work of entering the identifying data that in turn makes those images searchable. On Facebook, subscribers can tag photos of people in their network when they recognize them, simply by selecting the area of the photo in which they appear, clicking on the "Tag This Photo" command, and either choosing among their list of "Friends" or manually entering the name. (MySpace has a similar photo tagging feature.) All the photos in which a user is tagged appear in her Facebook profile, regardless of who took the pictures or posted them, and a link to those photos appears on the main page of a subscriber's profile (though Facebook does allow users to "untag" photos of themselves, in both their own photos and those posted by other users). While the process is made as simple as possible by these user-friendly software features, photo tagging is a manual and labor-intensive aspect of online photo management, performed unevenly and unsystematically. Some users are diligent about tagging their own photos when they post them online and tagging the untagged faces they recognize when browsing other people's albums. Others rarely bother to do this type of photo work, and millions of online photos remain untagged and thus unsearchable using text-based image retrieval techniques.

What facial recognition technology promises is content-based, as opposed to text-based, image retrieval. The effort to build online facial recognition search engines is part of the broader project of developing search techniques that use images themselves as search criteria, also known as content-based image retrieval, or CBIR. Generating a digital template from an image of a face and using it to search for a match in a database of photographed faces is a form of CBIR. CBIR is a difficult technical problem that computers are not able to perform very well, unless the search image and the target image are very similar in terms of lighting, pose, camera angle, and other visual details. (It works best when the two images are identical.) Still, the research area of CBIR has undergone tremendous growth since 2000. Research is increasingly focused on user intentions and expectations and the importance of building so-called human-centered image retrieval systems.[35] Image retrieval researchers are beginning to better understand the limitations of CBIR and the fact that robust image retrieval systems cannot rely exclusively on image content. There are two main reasons why: not only does the real world contain much more visual information than recorded images (the "sensory gap"), but even more important, images themselves can never contain the full range of possible human interpretations of those images (the "semantic gap"). According to the conclusion drawn in a recent survey of CBIR research, "The future of real-world image retrieval lies in exploiting both text and content-based search technologies."[36]

No one recognizes the need to combine both content- and text-based image retrieval techniques better than the search engine giant Google. At the main page of the Google Images search engine,[37] under the search bar, visitors are asked, "Want to help improve Google Image Search?" The question is followed by a link to the "Google Image Labeler" game, where willing players are randomly paired up with a partner to tag photos with whatever terms they can think of over a two-minute period, earning points when partners come up with the same descriptive terms. (Like a video game, the points are simply a measure of performance. They are not redeemable for cash or prizes of any sort.) "What do you need to participate?" asks Google. "Just an interest in helping Google improve the relevance of image search for users like yourself."[38] Google's image labeling game is a rather transparent effort to mine the vast pool of Internet users for free data-entry labor, appealing to site visitors to do their part in improving image search engines and the relevance of image search results. Polar Rose is another search engine that encourages Internet users to add metadata to images, specifically focusing on photos of faces: "With Polar Rose you can name the people you see in online photos and use our visual search engine to easily find more."[39]

As the people at Google and Polar Rose are well aware, successful automated image search techniques that produce "relevant" search results require a significant amount of human labor—that is, a special form of mental labor derived from humans' capacity to read the denotative and connotative content of images. For their part, experimental online applications of facial recognition technology encourage Internet users to participate in improving image retrieval techniques by bringing to the effort their highly developed social and cognitive capacity for facial recognition. Social networking and photo sharing sites are ideal laboratories for recruiting an army of well-trained "technicians" to engage in the tedious mental labor of photo tagging. In the case of MyFaceID, the application has a difficult time matching different images of the same person. What it does do adequately, however, is automatically locate faces posted in a profile and then compile them together on one screen, providing a more convenient way for users to tag those faces with their corresponding names. The aim is not to eliminate the need for photo tagging altogether, but to make manual tagging easier and more habitual, encouraging users to be more diligent about doing their part to ensure that the photos are more readily searchable. Text-based image retrieval is faster and more efficient that content-based searches (once the metadata has been added), and so the more photos tagged with identifying metadata the better the search results. The habituation of photo tagging among individual users greatly facilitates the searchability of images, involving Internet users in the massive archival project of managing the explosion of visual media content pouring onto the web.

Another way that developers of photo management applications promote and make productive use of the logic of new media interactivity is by encouraging users to experiment with the software prototypes and offer valuable feedback to developers.[40] Although developers had high hopes for them, early prototypes of online facial recognition search engine software did not work very well. Instructions to click on the "MyFaces" command of the MyFaceID application (the auto-tagging function) frequently resulted in a long pause followed by a "timeout" error message. The application was unable to successfully match two different photos of my face or any other faces in the photos I uploaded to my Facebook profile. For its part, the MyHeritage Celebrity Look Alikes application occasionally had difficulty finding faces in an image. For example, it could not locate the face of a person standing in front of a painting in a photo I uploaded, even though the person's face was directly facing the camera. When it was able to locate a face in an image, the Look Alikes application was reliable in that it would consistently come up with one or more celebrity matches, along with a numerical score representing the degree of likeness of each match. However, the program matched photos of my face with celebrities like Yoko Ono, Linda Carter, Katie Holmes, Renee Zellweger, and Debra Messing, to none of whom do I bear the slightest resemblance.

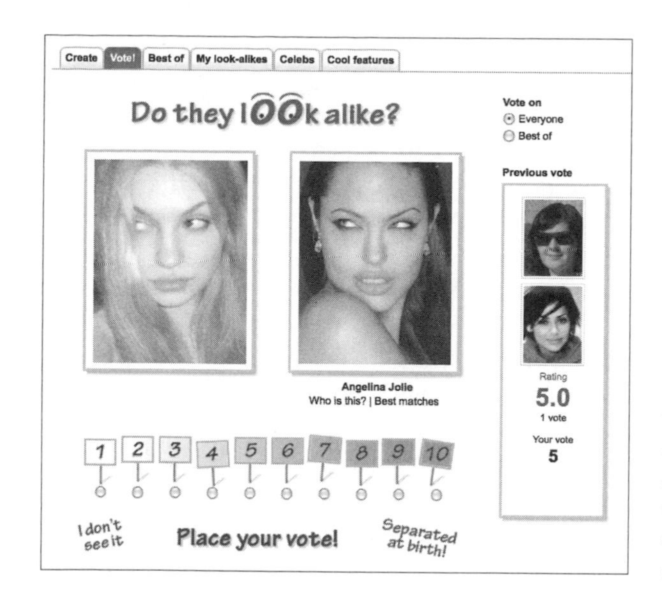


The MyHeritage Celebrity Look-Alikes application invites users to rate the quality of the facial match.

Although the limitations of the face matching software are worth noting, equally important is the way that the software developers acknowledged those limitations and invited user feedback. In the case of MyFaceID, users of Facebook were invited to participate in the development of the facial recognition search engine by experimenting with the application to identify photos of friends and strangers, with no real promise that the program would work that well. Users were not offered a fully functional and polished application, but instead one with kinks to be worked out and precise uses to be defined. On the MyFaceID information page on Facebook, the developers admitted that it is "still in development" but that they hoped it would help users "find new friends, organize your pictures, and get a bit of fun on the way."[41] The rough early stage of the software was confirmed by the awkwardly worded copy explaining the application to users, along with the small number of user reviews and wall postings at the application's Facebook page. Another explanation of the MyFaceID application told visitors that "this is a very early Beta version of our service," and users were invited to e-mail the company if they had a "feature suggestion" or noticed a bug.[42]

The MyHeritage Celebrity Look-Alikes application took a more defined form, and did not invite user comments. But the application did include a feature that allowed users to rate the celebrity resemblances generated by other users, suggesting that the program's results were of variable quality and inviting users to decide for themselves how well the program worked. Since the development of functioning automated facial recognition would depend fundamentally on getting computers to identify faces the way humans do (and doing an even better job at it), these user ratings provided valuable data for algorithm development and assessment. At the same time, the rating system encouraged a symbolic form of user participation in tech evaluation, involving users in a playful and seemingly innocuous form of experimentation and testing of new web software.[43]

These forms of user experimentation with photo management applications of facial recognition technology embody the dualistic, reflexive logic of "interactivity." User agency with respect to photo browsing and sharing practices becomes an object of scrutiny, so that knowledge of user behavior and preferences can be incorporated into design features, which in turn can be marketed back to those users as agency-enhancing tech "solutions." Developers gather feedback from people testing out the software in hopes of understanding precisely what users do, or want to do, with their online photo collections, what they are looking for in their image searches, and how facial recognition technology, given its limitations, can be adapted and optimized to produce the most marketable tech products. And while developers are constantly experimenting with new and more marketable applications, users are simultaneously encouraged to view their own participation and feedback as a form of self-experimentation, part of the conduct they are expected to engage in as a responsible and technically skilled citizens living and actively participating in a high-tech society. The idea that software developers would consider user feedback in their software design process casts them in a democratic light, portraying an image of themselves as carrying on the ethos of collaboration and collective decision making associated with the early development of the Internet. In turn, providing feedback about online software becomes a way for users to engage in good online behavior, envisioning themselves as active participants in the ongoing development of the World Wide Web.

## *Whose Space?*

In all its varied uses, online social networking is a new site of social regulation and government, where users offer up personal information about themselves to friends, strangers, and other interested parties, while simultaneously participating in monitoring the lives of others.[44] If displaying one's photos online is a way of exhibiting oneself to the world, then browsing, searching, and identifying the photos of others is a way of watching over them, a form of what Andrejevic refers to as "peer-to-peer" or "lateral surveillance."[45] There is nothing new about sharing one's photos with others, but the current online photo sharing boom, facilitated by the proliferation of digital cameras in a variety of forms, is further challenging the shifting boundary between public and private.[46] The act of publicly exhibiting private images, or merely sharing them with a select group of "friends," undoubtedly gives some Internet users a sense of satisfaction or empowerment. Revealing visual details about one's private life may even be viewed by some as a means of bucking conventional distinctions between public and private, a form of resistance to traditional ideas about what is appropriately hidden or kept secret.[47] But such public forms of self-disclosure are not empowering in any simple or straightforward sense. While people may have control over what photos they post online, they cannot control what photos of themselves are posted by other users, or what happens to their photos once they are in circulation. And the ever-present possibility of having personal photos appear on the Internet cannot help but change our sense of the boundaries around our private lives, and our sense of proprietorship and control over the full range of information that constitutes our identities. The effect is no doubt intensified the easier it becomes to locate and identify individuals in online image databases, a problem that extends both forward into the future as well as backward into the past, as any photograph ever taken has the potential to migrate onto the Internet.

In other words, if individuals find consumer applications of facial recognition technology useful for their own practices of self-expression and self-management, their adoption of the software for personal use also serves broader institutional aims. Online social networking and photo sharing sites are experimental test beds for CBIR research, and individual Internet users will not be the only beneficiaries of any improvements in image search techniques born of this experimentation. End users' experimentation with web applications of facial recognition technology is necessary in order to help developers understand how best to address its limitations and produce viable applications. Photo management and image retrieval are important problems not only for amateur photographers managing their photo collections. These are also pressing concerns for institutional users like military and law enforcement agencies, as well as news organizations, Hollywood editors, search engine developers, and other social actors with large image databases to manage. As we have seen in previous chapters, the business, law enforcement, and state security sectors have a vested interest in more sophisticated access control, data mining, and "smart surveillance" technologies. Viewing web applications of facial recognition technology narrowly in terms of the benefits or forms of "empowerment" they invest with individual users obfuscates the broader promise of the technology to improve the functionality and reach of large-scale surveillance and access control systems.

The contribution that experimental, "interactive" consumer applications can make to building better facial recognition systems is not limited to technical improvements in image retrieval techniques. Much like its uses for personal security, photo management applications help to introduce facial recognition technology into the self-fashioning activities of tech-savvy citizens, giving users a sense of technical mastery over the technology and in turn helping to alleviate the threat it seems to pose to our own ability to control what we do with our faces and our identities. Images of the face have long played a role in practices of self-identity formation, as well as in more formalized systems for the management and authentication of identity. The way certain people collect and use images of their faces and the faces of others is one of the techniques they use to build and manage their own personal biographies. Introducing facial recognition systems into these practices of the self plays a fundamental role in the institutionalization of the technology. What proponents rather deterministically refer to as "user acceptance" of biometric technologies is in fact a complicated matter of creating the kinds of people who will use them: people appropriately concerned with their personal security, watchful of themselves and others, and mindful of a compelling need to continuously experiment with new technologies as a matter of their own self-expression and self-improvement.

The question arises as to what happens to our understanding of the face itself, and face-to-face social interaction, as a result of these combined changes in the institutional and individual practices of facial image management. The applications of facial recognition technology examined here are designed to remedy problems arising from changes occurring in human social interaction along with the proliferation of disembodied identities. They promise to provide a technical solution to one of the central problems of mediated communication: "the possibility of interaction without personal or physical contact."[48] But facial recognition technology is limited in its capacity to resolve the lack of personal connection inherent in disembodied social interaction. This is because faces and facial images are not only markers of identity, they are also conveyers of meaning and affect, representing and misrepresenting the communicative aims and embodied affective expressions of their bearers. While facial recognition software may help identify faces and improve the searchability of facial image databases, the technology itself reveals nothing about the affective dimensions and subtle social cues that faces convey. In order for computers to "see" the face the way humans do, computers will need to do more than identify faces and distinguish them from one another. They will also need to discern the signifying expressions of the face—the aim of automated facial expression analysis. Where facial recognition technology denies the face its meaningful expressive qualities in order to transform it into an index of an individual identity, automated facial expression analysis treats the expressive movements of the face as a field of computational data about the varied meanings that those movements convey. In this way, it promises to bring the face and its expressive dimensions into the domain of human-computer interaction, so that human affective relations and human subjectivities themselves might be reinvented.

# 5

# Automated Facial Expression Analysis and the Mobilization of Affect

As he recalls it, Joseph Weizenbaum was moved to write his book *Computer Power and Human Reason* as a result of the public reaction to his experimental language analysis program, named ELIZA, that he developed at MIT in the mid-1960s.[1] ELIZA's incarnation as DOCTOR, a parody version of a psychotherapist asking inane reflexive questions, was wildly misinterpreted as an intelligent system, according to Weizenbaum. Much to his astonishment, some practicing psychiatrists actually thought that the DOCTOR program could be developed into a truly automated form of psychotherapy, "a therapeutic tool which can be made widely available to mental hospitals and psychiatric centers suffering a shortage of therapists."[2] The conscientious computer scientist objected:

> I had thought it essential, as a prerequisite of the very possibility that one person might help another learn to cope with his [*sic*] emotional problems, that the helper himself participate in the other's experience of those problems and, in large part by way of his own empathetic recognition of them, himself come to understand them. . . . That it was possible for even one practicing psychiatrist to advocate that this crucial component of the therapeutic process be entirely supplanted by pure technique—*that* I had not imagined![3]

Although others have since reinterpreted ELIZA in more techno-optimistic terms,[4] Weizenbaum's concern for his simulated psychiatrist's incapacity for "empathetic recognition" is instructive for a critical analysis of the effort to program computers to "see" the human face. It is especially relevant to a dimension of the research that I consider in this chapter. Whereas up until now I have focused largely on automated facial recognition as a technology of identification, here I investigate the development of automated facial *expression* analysis, the effort to program computers to recognize facial expressions as they form on and move across our faces. While facial recognition technology treats the face as a "blank somatic surface" to be differentiated from other faces as an index of identity, automated facial expression analysis treats the dynamic surface of the face as the site of differentiation.[5] The dimensions and intensities of facial movements are analyzed as indices of emotion and cognition, as a means of determining what people are thinking and feeling. Experiments in automated facial expression analysis—or AFEA, as I will refer to it in this chapter—represent a subfield of computer vision research, overlapping but often distinct from experiments in automated facial recognition.[6] The automation of facial expression analysis promises to accomplish what facial recognition technology fails to do: read the interior of the person off the surface of the face, using the face itself as a field of classifiable information about the individual.

The issues that Weizenbaum's program raised about the possibility of using computers to perform some of the intellectual labor of psychotherapy, as well as the ethical implications of doing so, arise again in the case of automated facial expression analysis. The project of AFEA is tightly bound to the field of psychology: psychological theories of facial expression and emotion inform the design of AFEA systems, and those systems in turn promise to advance the field's knowledge of facial expressions and emotion. Facial expression analysis is one approach to studying the psychophysiology of emotion, among an array of techniques for measuring the physiological manifestations of the affects. In this sense, there are important differences between the ELIZA program and the project of AFEA. For its part, ELIZA was not a computer vision program but a language analyzer designed to automatically manipulate text that a user consciously typed into a computer, the program drawing from a script to simulate the questions and responses of a psychotherapist. No one claimed that ELIZA could read the faces or emotions of the people with whom "she" was interacting. In fact, physiologists interested in emotion would have little use for ELIZA, long asserting "the poverty of human language to represent emotions" and a preference for examining affects in their pre-linguistic phase rather than relying on the emotional self-awareness of human subjects.[7] Like a long line of techniques before it, AFEA promises to bring this pre-linguistic phase of human emotion into greater visibility, irrespective of the conscious reflection of human subjects about their emotional states.

Beyond the methodological issues it raises for the psychological sciences, the idea that an automated form of facial expression analysis might replace in some cases the role of human beings as readers of one another's faces raises anew questions about the meaning, experience, and ethical substance of face-to-face versus mediated forms of communication. As John Durham Peters has argued, the concept of "communication" itself became intelligible as a result of the "disembodiment of interaction" associated with the rise of mass media like radio and television.[8] A similar claim might be made of the possibility of automated facial expression analysis: when the analysis of facial expressions of emotion becomes something that appears to happen independent of embodied human perception, human face-to-face affective relations take on new life as an object of inquiry. But just as believing that ELIZA could legitimately perform the role of psychotherapist required a willful blindness to the program's basic technical underpinnings and lack of capacity for empathy, believing that computers can be made to "see" facial expressions—and by extension the affective states of individuals—likewise involves forgetting or bracketing the layers of social construction and technical integration involved in the development of AFEA systems. Once again, getting computers to "see" the face involves building into these systems particular ways of seeing, rather than creating a technically neutral form of machine vision that is free from the ambiguities, imperfections, and subjective capacities of human visual perception. AFEA might simulate sight, but it cannot do so without a preprogrammed script.

Programming computers to identify facial expressions, especially under uncontrolled conditions, poses a difficult set of problems for computer scientists to tackle, in some ways even more challenging than programming computers to identify faces. Whereas facial recognition technology has already been integrated into some real-world surveillance and identification systems, AFEA is still in a more nascent stage of development in computer science. It has not been deployed on any significant scale, although there is at least one consumer application of automated *smile* recognition: Sony's Smile Shutter™ technology, integrated with some of its digital cameras, automatically snaps a photo when a person being photographed is smiling. This seemingly trivial application in fact points to the mutually constitutive relationship between the technology of photographic portraiture and the social conventions that govern facial expressions and their representation. Smiling became a regular feature of photo portraits only after shorter exposure times became possible, and the popular practice of "smiling for the camera" points to both the often *posed* status of facial expressions captured in images as well as the modern social expectation that people appear happy in images. Cameras with Smile Shutter™ technology incorporate within their design a historically and culturally specific injunction to represent oneself as perpetually cheerful, and the skill many of us have at displaying a happy face no doubt has some bearing on why the smile is the first expression computer programs are able to "see" with some reliability.

But beyond its capacity to provide automatic snapshots of the smiling face, AFEA remains largely confined to computer science laboratories and some basic research in "emotion science" (including one study investigating "the timing of the onset phase of spontaneous smiles").[9] According to a recent survey of the research, "The application of currently available automatic facial expression recognition systems is often very restricted due to the limited robustness and hard constraints imposed on the recording conditions."[10]

What we can say with certainty about automated facial expression analysis is that it represents an object of interest to both psychologists studying facial expressions and to computer scientists, including some of the same computer scientists developing facial recognition technology. It is also of interest to other social actors with a desire or perceived need to know more about what individuals are thinking and feeling beyond what they actually express in words. A chapter from Stan Li and Anil Jain's *Handbook of Face Recognition* offers a list of potential uses, including for the study of "emotion and paralinguistic communication," as well as more applied areas such as "clinical psychology, psychiatry, neurology, pain assessment, [and] lie detection."[11] The chapter also mentions those applications most compelling to computer scientists themselves: intelligent environments and multimodal human-computer interface, or HCI.[12] Other potential applications include monitoring the cognitive and emotional states of people in workplaces, airports, border control stations, and in the penal system, as well as for gathering affective and nonverbal data about consumers for market research purposes. If using AFEA to assess the emotional state of consumers in order to optimize their inducements to purchase seems like a stretch, consider that Disney has reportedly created a new, high-tech research laboratory that will allow scientists to "measure heart rate and skin conductivity and track the gaze of participants who are exposed to new ad models over the Internet, mobile devices and TV screens."[13] In fact, the application of AFEA for marketing purposes would be consistent with a history of marrying the methods of psychology to consumer research, dating back to the work of Paul Lazarsfeld at the Bureau for Applied Social Research in the 1950s and 1960s, and will likely represent one of the predominant uses should automated facial expression analysis take shape as a functioning technology.

While technical articles about AFEA typically include an introductory section that offers a list of potential applications, questions about the social and ethical implications of such technologies for human affective relations are almost never asked.[14] In fact, there is a concerted effort in the technical literature to distinguish facial expression analysis from the analysis of emotions, an insistence that they represent two distinct problems. For example, one technical article notes that "sometimes the facial expression analysis has been confused with emotion analysis in the computer vision domain. For emotion analysis, higher level knowledge is required."[15] Although the authors are right to be careful about making assumptions about the relationship between facial expressions and emotions, they nevertheless suggest that technical systems for the automatic interpretation of expression are possible through additional layers of knowledge. As we will see, this "higher level knowledge" is defined in terms of added layers of code, or more advanced ways of classifying facial measurements into discrete categories of emotion and emotional intensity. This move to separate the specific problem of facial expression recognition from the recognition of emotion defines the aims of any particular effort at computerization as narrowly as possible, separating it from the messy and exceedingly more difficult problems associated with the meaning of facial displays and their relationship to what people are thinking and feeling. In this way the research can facilitate the aim of automatically reading internal states off the face without claiming to provide a direct means of making such subjective determinations.

But regardless of how narrowly the technical experiments are defined, particular understandings and assumptions about what facial expressions are, what they mean, and what they can be made to do, inescapably have consequences for the development of automated facial expression analysis. The very possibility of automating the imperfect—and culturally and historically variable—human capacity to read the faces of others suggests that certain assumptions about facial expressions and their relationship to human affective relations are already in circulation, assumptions that posit affect as a physiological process capable of being not only coded and analyzed but also *engineered*. If inventing new technologies means inventing the kinds of people who will use them, as we saw in the previous chapter, human reinvention likewise involves designing new technologies specifically for that purpose.[16] As an instrument for automatically measuring, defining, and differentiating human faces and emotions, AFEA is envisioned as among an array of techniques of bodily measurement and visualization designed to measure and regulate affect, thereby providing a technical means for analyzing and remaking people. Like the psychological test, AFEA promises to be a highly adaptable technique, "a device for visualizing and inscribing individual difference in calculable form," thereby helping to create "calculable individuals" who can be encouraged or enjoined to reevaluate and reinvent themselves in response to their changing social and technological environments.[17]

## *Photography and the Expressive Face*

The idea that careful examination of the face and facial expressions could reveal something about the internal state, even the essential core of the individual, has a long history, one intimately tied to religious beliefs, the human sciences, and the development of photographic technologies. The science of physiognomy held legitimate and respected status in the Western world in the eighteenth and the nineteenth centuries, alleging to show evidence of connections between facial features and individual character. The prominent eighteenth-century promoter of physiognomy, Johann Kaspar Lavater, espoused the view that facial features were God-given signs of moral worth.[18] Lavater distinguished physiognomy from pathonomy; the former involved the study of the features as rest, while the latter concerned the study of facial expressions, or the "signs of the passions."[19] The Scottish physiologist Charles Bell similarly maintained the divine origin of facial traits, but thought that expressions, rather than the features at rest, were the true indicators of God's imprint on the face.[20]

Photography was used for the physiognomic and pathonomic analysis of the face virtually since its inception, invented at the height of "physiognomic culture" in the first half of the nineteenth century.[21] Physiognomic and pathonomic practitioners showed a special concern for reading insanity or dementia in both facial features and in the dynamics of facial expressions, using photography to capture the "face of madness." Hugh Welch Diamond, the founder of clinical psychiatric photography, used photographs as both a diagnostic tool as well as a method of treatment, "to present an indelible image of the patient for the patient's own study."[22] But no one seems to have made more productive use of the new medium of photography as a tool for studying facial expressions than the French neurologist Guillaume Duchenne. Duchenne shared his predecessor Charles Bell's belief in both the primacy of facial expressions as indicators of inherent qualities, and in their divine origin. Unlike many of his contemporaries, however, Duchenne was less concerned with understanding the facial typology of insanity, than with studying neurological dysfunction in the face, as well as "normal" mechanisms of facial movements.[23] Duchenne adopted not only the camera but also the new technology of electrical stimulation to induce expressions on the faces of his patients, staging bizarre photographic portraits, some in which he himself appears. The photographs, showing electrical instruments being applied to contorted faces, look frighteningly like experiments in torture, although apparently his subjects experienced only minor discomfort. As visual evidence of a fleeting moment in the shared history of photography and physiognomy, Duchenne's photos convey an attempt to capture and study the face in motion, along with the positivist scientific ambition to apply new technologies to make observable what theretofore had fallen below the threshold of knowledge.[24]

Where Duchenne saw the work of God in his portrait photographs, Charles Darwin used similar photos to challenge assumptions about the divine origins of facial features and expressions. Working in collaboration with the photographer Oscar Gustav Rejlander and drawing as well on Duchenne's photos, Darwin published *The Expression of Emotion in Man and Animals* in 1872, "one of the first scientific texts to use photographs as part of evidence for the theory being proposed."[25] It was a direct attack on the idea that facial expressions were put there by God, instead maintaining they were largely the vestiges of evolution, with evidence in continuities across cultures and even species.[26] The presence of published photos in Darwin's book no doubt lent it an air of cutting-edge research, the images working alongside his written argument to discredit the moral and religious claims of physiognomy. Darwin himself emphasized that he used only photos "made by the most instantaneous process" allowing for "the best means of observation."[27] But the technical constraints of one-half- to two-second exposure times meant that the images Darwin collected in fact required more sustained expressions—"in some cases, facial gestures created specifically for the purpose of photography and sold as 'character' images by commercial studios."[28] Whether placed there by God or by evolution, the fleeting, spontaneous nature of facial expressions often eluded photographic representation.

It is no surprise, then, that a fascination with capturing the human face in close-up informed the early development of motion picture photography, as Tom Gunning has shown.[29] Gunning refers to the celebrated potential of motion pictures to uncover new visual knowledge as the new medium's "gnostic impulse." For Béla Balázs and other utopian thinkers of the early twentieth century, cinematic technology was especially promising in its capacity to open up the face to new levels of detailed visual perception. Writing in the 1920s, Balázs emphasized the gnostic potential in the cinematic close-up of the human face: "It is the 'microphysiognomics' of the close-up that have given us this subtle play of feature, almost imperceptible yet also so convincing. The invisible face behind the visible has made its appearance."[30] The drive to make visible the dynamics of facial expressions at their most acute level of detail played a key role in the invention of motion picture technology and the early forms it took.

Since the mid-nineteenth century, scientists interested in the study of the human face made productive use of the new photographic technologies, whatever their religious or scientific persuasion. Still photography, and later motion pictures, invigorated their work and redefined their methods, creating a type of archive—a source of observable, classifiable, visual information—that had not existed before. And the expanding archive of facial representations no doubt allowed the face itself to be reenvisioned as a sort of visual archive—a field of newly detailed, classifiable information about human beings. The medium of photography emerged and developed in intimate relationship with historical transformations in what were considered legitimate, properly scientific modes of classification and knowledge production, including methods for interpreting the face and facial expressions. The imprecise science of physiognomy, heavily inflected with religion and moral judgments, was supplanted with a seemingly more meticulous, methodical, and positivist form of facial analysis, one whose claim to truth was based in scientific and technical rather than theological foundations. Of course, this is not to say that there was any less concern with establishing the relationship between the exterior of the face and the essential interior qualities of the person.[31] The imprint of God may have been lifted, but the marks of nature remained and even intensified.

The history of developing and applying photography for the study of the face and facial expressions demonstrates the ways in which the face itself, or at least the way we *see* the face, is imbued with historical meaning. In other words, this history suggests that there is no essential link between facial expressions and what humans think or feel, nor is there any essential way in which humans recognize and interpret facial displays; there are only different ways of *conceptualizing* what facial expressions are and what they mean, and these understandings have implications for the human "emotional range" in any given social and historical context, affecting how human beings comport and regulate their emotions and how they interpret the emotional behaviors of others. In his history of sound reproduction, Jonathan Sterne makes a similar point about sound and hearing. He writes, "It is impossible to 'merely describe' the faculty of hearing in its natural state [because] . . . the language we use to describe sound and hearing comes weighted down with decades or centuries of cultural baggage."[32] Like sound and hearing, facial expressions, and our ways of seeing and interpreting them, "rest at the in-between point of nature and culture."[33] Photographic technologies have been developed and applied to define the categories of facial expression and to link those categories to variable states of human emotion. But expressions and emotions are far from purely physiological processes that exist in nature waiting to be discovered. Affective behaviors have a deeply relational basis, and one that has altered over time in tandem with changing forms of social organization.[34]

Although the scientists working on facial expression research may have had a narrow range of interests and motives, their research derived from and contributed to broader transformations in the social norms governing the expression of emotion, providing a more technical and scientific framework for the social regulation of affective behaviors. Affective behavior has been the target of social regulation over a long historical period in the West, bound up with transformations in forms of social organization associated with modernization. Norbert Elias has examined at length this "civilizing process," the long-term changes in the affect and control structures of people living in Western societies.[35] This so-called civilizing process occurred in complex relationship with the movement toward higher levels of social differentiation and integration, as people formed lengthening chains of interdependence. Key to the "civilizing process" was the consolidation of control over the use of physical force by states, enabling the formation of pacified social spaces in which people could live and interact in relative safety. Within these secure, socially regulated spaces, self-control over affective conduct became de rigueur.[36] Increasingly minute gestures and details of conduct came to have greater meaning, and a high level of self-control over affective behaviors became "second nature" for "civilized" individuals, imprinted on them from an early age.[37] Far from existing in some pure biological form in the body, emotions like shame, fear, and guilt—and especially "man's [*sic*] fear of himself, of being overcome by his own affective impulses"—"emerged in social interaction with others, directly or indirectly induced in a person by other people."[38] According to Elias, the strength of affective controls, as well as their form and the role they play in shaping individual personalities, depends fundamentally on the structure of societies and the place and fate of specific individuals within those societies.[39]

Proof of the more broad-based social-regulatory implications of the study of affect can be found in the sheer range of techniques devised to study it. The application of photography to the study of facial expressions was only one among an array of new techniques that scientists began to employ in the late nineteenth and early twentieth centuries in order to transform emotion into an object of scientific knowledge—something that could be studied in a laboratory. Instruments that measured mundane human physiological processes, like the pneumograph, kymograph, sphygmomanometer, galvanometer, and thermometer, were put to the task of measuring and visualizing emotion, locating affect in the bodies of research subjects and exteriorizing it in visual representations.[40] As Otniel Dror has written, the new "techno-representations" generated by these instruments were used to displace emotion from interior to exterior, "from the privacy of the mind to the communal space of representation, from personal experience to scientific knowledge, and from the subjective to the objective."[41] According to Dror, it was the scientists of emotion themselves, performing experiments on their own bodies in the laboratory and demonstrating their emotional control and agility, who were perhaps most directly affected by the scientific study and objectification of emotion.[42] Their experiments led to "a modern and particularly masculine form of communicating and exchanging emotion" in a scientific setting, creating "an emotional economy in the laboratory."[43] But their research had implications well beyond the laboratory, as the mechanically rendered physiological manifestations of emotion helped redefine what it meant to feel, what exactly emotions were, and how they could be differentiated and manipulated. The scientific study of the physiological manifestations of emotion added a technical, instrumental dimension to the social regulation of affect.

The history of scientific interest in facial expressions elucidates not only the interpretative flexibility of the face over the time, but also the ways in which the photographic-scientific analysis of the face was, and continues to be, bound up with what Ian Hacking calls "making up people," creating categories of human identity that change "the space of possibilities for personhood."[44] Focusing on the official statistics of the nineteenth century, Hacking argues that different kinds of human beings and human actions came into being hand-in-hand with the invention of categories for labeling people and their behaviors.[45] Counting and classifying types of suicides, for example, helped bring into existence the phenomenon of suicide itself and the range of motives for it. People may very well have killed themselves before the statistical study of suicide, but the labeling and differentiation of the act and those who committed it opened up new possibilities for defining it and even performing it: "The systems of reporting positively created an entire ethos of suicide, right down to the suicide note, an art form that previously was virtually unknown apart from the rare noble suicide of state."[46]

The psychological sciences that emerged to analyze, explain, and treat such maladies of the mind have played a special role in making up the kinds of persons that modern subjects take themselves to be.[47] And as a subfield of the modern disciplines of psychology and psychophysiology, the science of facial expression analysis has its own unique role to play in making up people. Facial expression analysis has not involved a simple and straightforward discovery of the details and dynamics of expressions as they exist on every human face, universally across cultures and throughout history. Instead, the science of facial expression has worked in tandem with other knowledges of affect to define the normal range of emotional responses to the world and the ways those responses are enacted through the face and the body.

The relationship between the study of facial expression and the social regulation of affect continues to the present day, albeit not in a continuous line. Darwin's evolutionary explanation of facial expressions eventually found renewed interest in the work of psychologists like Silvan Tomkins, Paul Ekman, and Carroll Izard in the 1960s and 1970s. These scholars brought the study of emotion, and its expression in the face, back to the forefront of psychology after a long period—Erika Rosenberg labels it the "zeitgeist of behaviorism"—in which emotion was steadfastly rejected as an "unobservable."[48] Today, Ekman's neo-Darwinian theory of facial affect dominates the psychological study of facial expressions, but his influence extends beyond the academic field of psychology. In 2008, his work on deception detection became the inspiration behind a Fox television show, *Lie to Me*, about a fictional team of expert truth consultants with a disturbingly perfect track record of catching liars. In real life, *Time* magazine named Ekman one of the world's one hundred most influential people in 2009.[49]

In his early work, Ekman claimed to find support for the Darwinian position that a certain class of facial expressions was universal across culture and history, a position that now finds broad, though not unanimous or unqualified, support among psychologists of emotion.[50] Establishing the universality of facial expressions was especially important to both the reliability and validity of the work that followed—the project, on the part of Ekman and his colleague Wallace Friesen, to create a comprehensive facial expression classification scheme that could be applied to every human face, and that could be used to make discoveries about facial affect that could be generalized universally—in other words, a method for making discoveries about facial expressions and their relationship to felt emotions that would apply to all human beings. According to their own accounts, Ekman and Friesen developed their coding system in the 1970s, spending eight years studying facial expressions in order to differentiate every visually distinguishable facial movement.[51] As a result of this work, much of which was performed on their own faces, they created a schema of forty-four discrete facial "action units" (AUs), individual muscle movements combinable to form many of the different possible facial displays, some related to emotions and some not. The resulting "Facial Action Coding System," or FACS, became the "gold standard" measurement system for facial expressions. Although performed manually by trained human coders, FACS was, from the beginning, a computational model of facial expression, and as we will see, it promises to play a central role in the development of automated facial expression analysis.

## *The Problem of Automated Facial Expression Analysis*

Although photography has been used to analyze facial expressions virtually since its inception, the automation of facial expression recognition and interpretation is a relatively new and interdisciplinary endeavor involving various subfields within computer science, cognitive science, and psychology. Some of the challenges involved in creating computer programs that can recognize facial expressions became evident in an early study on "pattern recognition of human emotion expression," presented at the Fourth International Joint Conference on Pattern Recognition in Kyoto, Japan, in 1978.[52] To conduct their study, the researchers made films of actors performing the expressions, with coordinate marks placed on their faces based on the anatomical advice of a medical doctor. The actors were instructed to portray a variety of fanciful situations to convey happiness, anger, fear, grief, and jealousy, after which the researchers selected 358 frames from the films to produce enlarged still prints for analysis. The study's findings, published in the conference proceedings, reported that the program was able to recognize the expression of happiness fairly well but had a more difficult time with anger and jealousy. The movement of the head posed complications for the tracking of marked coordinates on the face, making it difficult to discern head motion from facial muscle movements and presaging the need for face tracking and feature extraction techniques. The study also included a comparison of the pattern recognition technique with an experiment in which human subjects judged the expressions of the filmed actors, finding "some similarity" in the "visual recognition rate" between humans and the computer program.[53] The researchers envisioned pattern recognition of facial displays as applicable to "the diagnosis of a psychiatric patient whose symptom is often seen in his facial expressions," and recommended further research that would identify the facial parameters most relevant for identifying different expressions, along with comparisons of "parameters between those of normal persons and those from patients."[54]

As this account suggests, the earliest efforts to develop methods of computerized facial expression analysis envisioned the technology as useful for psychiatric applications, especially for differentiating normal from pathological patients. Insofar as pattern recognition techniques were viewed as a potential means of automatically identifying facial expressions of emotions in place of the human capacity to do so, such techniques shared a common bond with ELIZA, Weizenbaum's language analysis program, which some practicing psychiatrists thought might function as an automated form of therapy for underserved patients. Of course, no one was explicitly suggesting that automated facial analysis techniques might entirely fulfill the role of psychiatrist. The promise seemed to lay more in the precise identification and differentiation of facial pathology, using computerized video analysis to visualize and make obvious what was imperceptible or ambiguous even to the trained eyes of experts. Computational models held out the promise of more accurate and meticulous recognition of facial expressions, free of the subjective judgments and imperfect modes of interpretation inherent in human perception.

But paradoxically, in order for computer vision to be deemed similar to or better than that of humans, the automation of facial expression analysis would require the comparison of computerized techniques against human perceptual capacities. In other words, an "accurate" reading of facial expressions depended not so much on what the expressing subject felt, but more so on what other human beings saw when they looked at an expressive face. However imperfect humans were at reading facial expressions, those expressions only came to have meaning in humans' interactions with one another. Whether computers could in fact be programmed to identify facial expressions with the same or better "visual recognition rate" as humans would depend on translating the messy ambiguity of human face perception and face-to-face interaction into more exacting computational models. And whether facial expressions and their interpretation were amenable to an information processing model of communication was by no means assured.

Not surprisingly, programming computers to recognize facial expressions, in fully automatic and real-time fashion, has posed a host of technical challenges. Some of these challenges no doubt also cause problems for humans, but they create unique difficulties for computer algorithms. Problems include dealing with differences in facial appearances among populations and individual differences in expressiveness, as well as discerning transitions among expressions, intensities of expression, and deliberate versus spontaneous expressions.[55] (Like Darwin's "character" images, most of the research on AFEA has relied on deliberate or posed rather than spontaneous expressions.) While facial identification requires computers to determine what is unique about a face, facial expression recognition requires computers to disregard what is unique, or at least to differentiate the face in a different way. Facial expressions are not static but involve faces in motion, and getting computers to parse different facial movements from a stream of behavior is a daunting computational problem. Head orientation and scene complexity add to the challenge, and image acquisition, quality, and resolution issues likewise introduce enormous complications that make real-time AFEA an especially difficult problem for computer vision. Conditions that most humans manage with relative ease become major complications of information processing. A recent technical description of AFEA notes some of these complications:

> Within an image sequence, changes in head position relative to the light source and variation in ambient lighting have potentially significant effects on face expression analysis. A light source above the subject's head causes shadows to fall between the brows, which can obscure the eyes, especially for subjects with pronounced bone structure or hair. Methods that work well in studio lighting may perform poorly in more natural lighting (e.g., through an exterior window) when the angle of lighting changes across an image sequence.[56]

Humans interpret one another's faces in highly imperfect ways based on social conventions that vary culturally and historically. But while there may be many factors complicating the way people convey and interpret facial expressions in any given interaction, the radically different colors and textures that different lighting can produce when it comes in contact with the face is not usually a major confounding issue (unless of course one happens to be a visual artist, in which case one might pay keen attention to these variations). Dim lighting can sometimes present challenges for human perception, but variations in lighting above a certain level of visibility hardly register at all for our ability to see facial expressions. In general, although not always particularly adept at understanding one another's expressions, most people are capable of easily ignoring complications introduced by lighting or other non-signifying variations in the appearance of faces. Most humans learn social conventions for attending to and disregarding certain kinds of variations in facial appearances, and for ignoring inadvertent facial movements and other gestures as necessary to avoid awkward interactions. In other words, these forms of interaction become a sort of "second nature."[57] In his famous essay "On Face-Work," Erving Goffman addresses this process at the length. Goffman defines "face" not as a specific part of the body but as the "positive social value" individuals effectively claim for themselves in interpersonal encounters, "an image of self delineated in terms of approved social attributes."[58] In many cultures, Goffman maintains, face-to-face encounters carry an implicit expectation that the parties involved will have a measure of respect for one another and sustain a standard of considerateness. In any particular encounter with others, a person "is expected to go to certain lengths to save the feelings and the face of others present, and . . . to do this willingly and spontaneously because of emotional identification with the others and with their feelings."[59] This involves subtle determinations about what to pay attention to and what to ignore in the actions and expressions of others. While at times it is important to pay close attention to the facial signals of others, at other moments it may be more appropriate to avert one's gaze or simply not acknowledge certain expressions, especially if they suggest felt emotions that the bearer may not intend to reveal.

The problem of how to train computers to make judgments about what to focus on and what to ignore in any given complex scene is not unique to facial expression analysis, but is instead a more general problem in the research and development of artificial intelligence and automated perception. As Hubert Dreyfus has argued in his philosophical critique of artificial intelligence, computerized pattern recognition techniques lack key characteristics of human perception and intelligence. More specifically, two dimensions of intelligence these techniques do not possess are "fringe consciousness" and "context-dependent ambiguity tolerance."[60] Our "fringe consciousness" is the background or marginal forms of awareness that enable us to concentrate on relevant information. Dreyfus uses the example of "our vague awareness of the faces in a crowd when we search for a friend" to demonstrate "the power of the fringes of consciousness to concentrate information concerning our peripheral experience."[61] Equally if not more important for seeing and interpreting facial expressions, and for human affective relations in general, is the human capacity for ambiguity tolerance—"our ability to narrow down the spectrum of possible meanings by ignoring what, out of context, would be ambiguities."[62]

Although Dreyfus's critique of AI is instructive for understanding the challenges of AFEA, his analysis makes the implicit and problematic assumption that dimensions of human intelligence are universal and transhistorical—unchanging characteristics that all humans have exhibited at all times in all societies. As we have seen, Norbert Elias has made the compelling case that human affective relations change in relationship to changing forms of social organization, and particular ways of expressing and controlling affects develop over long historical periods. From this perspective, the problem is not that computers have a difficult time gaining or exhibiting the essential, universal characteristics of human perception and intelligence. Instead, the challenge lies in designing computational models of perception that can appear to do what humans do as "second nature," or by acquired social convention. It is in fact the case that facial expressions—their relationship to emotion and cognition and the functions they perform in face-to-face relations—are both context-dependent and full of ambiguity, and so require an enormous level of ambiguity tolerance, as well as the ability to generalize across contexts using forms of "common sense." But there are no doubt wide historical and cultural variations in the sorts of expressions and gestures considered to be ambiguous, along with disparities in levels of expected tolerance for such ambiguities. It is likewise highly questionable whether humans possess a universal form of common sense. The kinds of practical judgments people make based on experience are more adequately understood as being shaped by a complex, culturally specific, and historically contingent range of social norms and expectations.

It is no wonder that computers have a difficult time "seeing" human faces the way humans do. The problem of how to program computers to make determinations about what to pay attention to and what to ignore in facial movements and appearances is not simply a matter of measuring a physiological process, or of controlling for lighting conditions and other issues of image variability, however complex these issues are for computational models of vision. Defined narrowly in these terms, the problem would seem to simply require more computer-processing power, or a more sophisticated computational approach. This is often the assumption that computer scientists make. For example, insofar as they metaphorically function like neurons, neural networks—or "networks of simple interconnected processing units that operate in parallel"—are seen as offering a promising possibility for addressing the challenges of automating facial expression analysis.[63] But it remains the case with neural networks that "unless the class of possible generalizations is restricted in an appropriate a priori manner, nothing resembling human generalizations can be confidently expected," as Dreyfus notes.[64] If a designer of a system "restricts the [neural] network to a predefined class of appropriate responses, the network will be exhibiting the intelligence built into it by the designer for that context but will not have the common sense that would enable it to adapt to other contexts."[65] Given that facial expressions are fundamentally ambiguous and context-dependent, the difficulties computers have in dealing with ambiguity and in generalizing across contexts are especially relevant to the prospect of developing functioning AFEA systems. The kinds of choices that computerized facial expression analysis must incorporate in order to "see" the face the way humans do are choices fundamentally shaped by social convention. In other words, no form of AFEA could encompass a universal, transhistorical form of face perception. Instead, any functioning AFEA system would necessarily incorporate and exhibit particular *ways of seeing* the face and facial expressions.

By highlighting some of the limitations of a computational model of facial expression recognition, I do not mean to argue that accomplishing functioning AFEA is impossible, that computers will never adequately "see" facial expressions, or that there are no physiological dimensions to emotion. Rather, I want to emphasize that the very possibility of AFEA raises critical issues about the nature of emotion and human affective relations, especially insofar as the computational model becomes an instrument for conceptualizing what emotions are and how they work. Just as the turn to biometric identification signals a redefinition of identity as disembodied data that circulates over networks, the prospect of AFEA similarly points to reconceptualization of face-to-face, affective relations to incorporate a computational model of face perception. AFEA is being envisioned for a range of applications, from basic research in psychology, diagnostic and therapeutic applications to market research, deception detection, and human-computer interfaces. This makes it especially important to understand how the computational model of facial expression interpretation works—including the classification or coding system on which it is based—and to consider the potential effects of computational models of affect on the social conventions that govern the way human beings interpret facial expressions and communicate with one another. "Technologies are associated with habits and practices," writes Jonathan Sterne, "sometimes crystallizing them and sometimes promoting them. They are structured by human practices so that they may in turn structure human practices."[66]

## *Coding Facial Expressions*

The problem of programming computers to "see" facial expressions has received sustained research attention since the 1990s, along with advancements in related techniques for the computational analysis of the face, like face detection, face tracking, and feature extraction.[67] Computer scientists have approached the problem of AFEA using a range of methods, generally divided into two categories: judgment-based and sign-based approaches.[68] Many of the automated facial expression analysis systems developed in the 1990s, like the early research in the 1970s, used judgment-based methods, meaning they would "attempt to recognize a small set of prototypic emotional expressions" using images of the whole face.[69] A consensus seemed to emerge rather quickly, however, that judgment-based approaches were too limited for many applications, since people rarely portray the prototypical expressions in everyday life, and the full range of facial muscle movements conveyed much more information, and often more subtle messages, than those communicated by obvious displays of delineated and discretely defined emotions like disgust, fear, joy, surprise, sadness, and anger. Computer scientists began to explore how to devise sign-based, or "sign vehicle–based" approaches, born out of the Shannon-Weaver mathematical theory of communication, which held that "the best strategy to measure signals is to measure the sign vehicles on which they are based."[70] Sign vehicle–based approaches would ostensibly capture more complete information from the face by eschewing judgment about the meaning of expressions, and instead treating the face as pure information, coding facial motion and deformation into a predetermined set of visual classes.

The Facial Action Coding System seemed to provide precisely what computer scientists needed to create automated systems that would be able to differentiate the dimensions of facial displays beyond a series of prototypical expressions. FACS offered a classification system for facial expressions that, although designed to be performed manually by trained human coders, was nevertheless based on the sign vehicle or information processing approach to communication. In other words, FACS was already a computational model for differentiating the dimensions of visible facial motion.[71] In contrast to other investigations of facial expressions that relied on inferences that observers drew about facial expressions, or studies that devised methods to describe a particular sample of facial behavior, Ekman and Friesen explain that their goal in developing FACS was to invent a comprehensive system that could identify and distinguish every visible facial movement, "free of any theoretical bias about the possible meaning of facial behaviors," and so would be applicable to a wide range of studies, including those unrelated to emotion.[72] They also wanted to devise an unobtrusive system that could be performed on people who were unaware that their faces where being scrutinized, unlike electromyographic (EMG) measurements, which involved attaching electrodes to the face. Ekman and Friesen believed, rightly so, that knowledge of being monitored would lead subjects to alter their facial behavior.[73]

Whereas computer scientists saw the Facial Action Coding System as a promising basis on which to build automated facial expression recognition systems, Ekman and other psychologists saw new computer processing techniques as offering promising possibilities for the automation of FACS. A major impediment to the widespread use of FACS in the psychological sciences and other contexts, according to psychologists, "is the time required both to train human experts and to manually score the video tape."[74] Again and again the psychological literature dealing with FACS notes that "the time-intensive process of observational facial measurement limits the rate at which new research can be completed."[75] To achieve even minimal competency in FACS, it takes a human being over one hundred hours of training, and one minute of videotape takes about an hour to code manually. Not only is the process time- and labor-intensive, but "human-observer-based methods" of coding the face are seen as too variable, incorporating too much error: "They are difficult to standardize, especially across laboratories and over time."[76] In other words, the way trained humans coders interpret the meaning of facial actions—let alone laypeople untrained in precise facial coding—is viewed as too subjective and tainted with inaccuracies.

In response to the labor problems associated with the manual coding of FACS, Ekman and a group of researchers in the Machine Perception Laboratory at the Salk Institute at University of California, San Diego, began a project to automate the coding system beginning in the early 1990s. The project, funded by the National Science Foundation, involved the development of a neural network system that would automate the time-consuming and expensive process of coding facial expressions. According to the researchers at UCSD, automating FACS "would not only increase the speed of coding, it would also improve the reliability, precision, and temporal resolution of facial measurement."[77] In addition, an automated system promised to "make facial expression measurement more widely accessible as a research tool in behavioral science and investigations of the neural substrates of emotion"—in other words, it would facilitate research that aimed to connect facial expressions to brain function as a way of explaining emotion physiologically.[78] As part of its support for the initiative, the NSF also sponsored a workshop that brought together psychologists and computer vision scientists from other institutions interested in the problem, an event that spurred stepped-up research activity and led to a series of annual international conferences on automated face and gesture recognition.[79] Scientists at institutions like Carnegie Mellon's Robotics Institute and the Beckman Institute at University of Illinois, Urbana-Champaign, joined the effort to build FACS-based AFEA systems. The interest of Ekman and his colleagues in automating their classification system in order to make it faster and more accurate converged with the interests of computer scientists in developing automated forms of facial expression recognition in order to create more sophisticated forms of human-computer interaction.

Since the Facial Action Coding System is seen as the most promising sign-based classification system for AFEA, FACS itself deserves closer attention. According to their own accounts, Ekman and Friesen did much of the research to develop FACS on their own faces, using anatomy texts (and occasionally electrical stimulation) to train themselves to fire specific facial muscles, along with mirrors and cameras to capture and identify the visible muscle movements. They assigned each of the forty-four action units they identified with a number and a name, such as "Inner Brow Raiser" or "Lip Corner Puller." The action units are meant to measure the observable surface movements, not to correspond precisely with particular muscles: some single action units include more than one muscle movement, and some muscles are involved in more than one action unit. People learn FACS and become experts by studying the FACS manual and its accompanying videos, available for purchase online, and then taking a standardized test. Trained human coders perform the labor of classification by viewing video and filmed images that they can stop, rewind, and view in slow motion. Scoring involves identifying what facial actions occurred as well as their temporal locations—not only what happens on the face but also when each action begins and ends, coded by video frame. Coders also score the intensity of expression on a selection of the action units that Ekman and his colleagues determined to be important indicators of emotional intensity, and they code whether particular facial actions are asymmetrical (more pronounced on one side of the face) or unilateral (an action on only one side of the face).[80]

Although FACS is a standardized system, human coders do not always agree about what they see, hardly surprising given the range of variation in facial appearances and ways of interpreting them. The problem of "intercoder agreement" in analyses of visible facial actions receives special attention in the FACS literature. Studies that make use of FACS ideally employ more than one human coder to score the faces of human subjects, and the rate of agreement among coders is reported as a measure of the reliability of research findings.[81] Whether FACS represents a "reliable" system of facial expression measurement is an issue of special concern to both psychologists and computer scientists.

This concern for consistency and replicability of facial action coding is revealing of the way in which the accuracy of FACS is constructed. Maximizing the level of agreement among human coders concerning the appearance, timing, and intensity of facial actions becomes a means of establishing the accuracy or "ground truth" of facial coding for AFEA algorithm development. Standardization is equated with accuracy and truth; if the facial coding system can provide consistent results no matter who is doing the coding, what specific faces are being coded, or in what context, then the method gains credibility as an accurate measure of reality.

Paradoxically, while *automating* FACS is seen as a way of further standardizing the classification system and eliminating human errors in the coding of facial actions, the successful development of AFEA depends on the reliability of *manual* coding. "The reliability of manually labeled images is a critical concern for machine learning algorithms," as one scientist notes, because "if ground truth is contaminated by 20–30% error, which is not uncommon, that is a significant drag on algorithm performance."[82] Here again we see that the accuracy of both FACS and AFEA is constructed in part through the process of standardization itself. The more people and computers can be made to agree about facial actions and intensities (as determined by FACS), the more accurate the classification system is said to be. This means that the "accuracy" of the system is defined not in terms of a correspondence between the system and a person whose face is being analyzed (or the full range of ways a facial expression might be interpreted), but in terms of the system's internal agreement—that is, whether the classification system consistently agrees with itself. While human perception is important to the design of both FACS and FACS-based AFEA, what is equally important is that both human coders and automated systems are trained to consistently identify the discrete facial action units of the standardized Facial Action Coding System—in other words, that the system functions as a closed loop that can operate with minimal room for "error" or unique interpretations of facial expressions.

The real power of FACS comes into play when this form of "accuracy"—deriving from the process of standardization—in turn gets pushed back out onto faces, as the standardized coding system comes to define what facial expressions mean and how they work. Facial expressions are made to fit the technology and the decontextualized, mathematical theory of information that informs its design. In her discussion of the Shannon-Weaver mathematical model of information, Katherine Hayles makes a similar point.[83] She explains that, for practical reasons, Shannon's theory purposely decontextualized information, divorcing it from meaning and disregarding the mind-sets of individuals involved in communicative exchanges. In this way, information would have a stable value as it migrated from one context to another, rather than having to be revalued as its meaning changed from context to context. Divorcing information from meaning fulfilled the instrumental purpose of designing efficient communication networks. But in turn, the engineering model of information became a dominant way of conceptualizing what information was, in all of its widely varying forms: "A simplification necessitated by engineering considerations [became] an ideology in which a reified concept of information [was] treated as if it were fully commensurate with the complexities of human thought."[84]

If a decontextualized model of information cannot fully represent human thought, it is likewise questionable whether it can capture the full complexity of facial expressions, emotions, and affective relations. FACS divorces facial actions from their meaning for instrumental reasons—in order to make facial expressions fully calculable. But at some point those expressions have to be invested with meaning, translated back into meaningful interpretations of emotion. Psychologists studying facial expressions are fully aware that agreement among coders about the appearance of certain facial actions does not mean that those actions are a valid measure of emotion. Even a perfect level of intercoder agreement cannot bind particular facial actions to particular emotions. At some point, a higher level of interpretation of facial action coding is required to make sense of the codes, to link combinations of action units to interpretations of their emotional, cognitive, or physiological associations. Ekman and Friesen addressed the question of the validity of FACS as an "emotion-signal" system in the earliest reports of their work, giving special attention to the need "to have some criterion independent of FACS to determine just which emotion, at what intensity, is being experienced at any given moment by a particular person."[85] They began to translate the interpretation of facial actions into another classification or coding system, following up FACS with efforts to develop emotion-coding schemes associated with FACS scoring.

The more interpretive system that Ekman and his colleagues began to develop sought to extend the standardized, information theoretical–approach to a decidedly more interpretive level of analysis. This interpretive emotion-coding system, initially called "Emotion FACS" or "EMFACS," focused on a certain set of "core AUs"—muscle actions and combinations of actions that Ekman and his colleagues found to be associated with particular expressions of emotion.[86] After devising a preliminary emotion-coding system, they built a relational database for the EMFACS data, and renamed it the Facial Action Coding System Affect Interpretation Dictionary, or FACSAID, a classification system designed to match "facial events with emotion events coded from previous studies."[87] The FACSAID database was designed to contain several types of data: "representations of facial expressions in terms of FACS, representations of meanings that can be attached to facial behaviors, and other facts about the behaviors, such as who attached meaning to a facial expression, how many times the behavior has been observed, pictorial representations of the behavior, etc."[88] An online description of FACSAID acknowledges that "one of the weaknesses in the database in its current form is that the rules for interpreting behaviors are not explicit, the authority of the experts who interpreted them being the only basis for confidence in their accuracy."[89] It also notes that "this problem will be addressed" in the next version of the FACSAID database, suggesting that the rules of interpretation will be provided to database users attempting to make sense of FACS scores. Whether and how the new database will represent disputes over facial expression interpretation, and how it will otherwise incorporate ambiguity, is not offered in this particular explanation of FACSAID.[90]

As systems for classifying facial actions and their relationships to emotions, FACS, EMFACS, and FACSAID are very much in keeping with Bowker and Star's argument that classification systems represent sites of political and ethical work, and that they perform a sort of reverse engineering on the world they aim to classify.[91] FACS itself represents an effort to differentiate, multiply, and limit the meaning of the face and its movements, to establish a new language and new techniques that rationalize variable facial appearances. Adopting the model of information theory, FACS constructs a framework whereby the face is translated into signals that can be divorced from their meaning, as disembodied entities that exist separately from "how they are understood by their receiver, or the intention behind their transmission."[92] While FACS itself is meant to be purely descriptive, the action units are scored in combinations in order to represent the more complex and meaningful facial displays of emotion.[93] The accuracy and precision of EMFACS or FACSAID as an *emotion*-signal system is constructed in part through the claims to objectivity associated with the initial stage of FACS coding, where the face is treated as pure information.[94]

This move to strip facial movements of their content and reduce them to a sort of informational essence is a way of investing FACS with "computational objectivity," an updated form of the "mechanical objectivity" associated with earlier photographic and visualizing technologies.[95] As Daston and Galison have elaborated, early techniques of mechanical reproduction held out the promise of judgment-free representation, of "images uncontaminated by interpretation."[96] But cameras and other measuring instruments never made good on their promise to fulfill the scientific-moral imperative of banning interpretation and eliminating the biasing presence of human observers. FACS likewise only temporarily brackets an interpretive framework for facial expression analysis. The purpose of the classification scheme is not to simply measure facial motion for its own sake, but to enable psychologists and other users of the facial coding system to make new *meaning* of the face—to examine and map out the surface movements, temporal dimensions, and intensities of the face with the aim of interpreting the interior state of the person, claiming that those surface movements can provide direct, unmediated access to embodied, felt emotion.

But it must be emphasized that Ekman and Friesen's classification system, the most promising basis at the moment for the development of AFEA, does not capture the essential, material features of facial movements that exist on every human face. It does not delineate a set of facial expression components placed there by God, evolution, nature, or culture, waiting for psychologists and computer scientists to lift them off the face and place them on a grid so that their true meanings can be established scientifically. Instead, FACS defines the components of facial expressions in the process of classifying them, enabling the production of newly detailed knowledge about the face. The coding system differentiates facial phenomena, opening up faces and all their subtleties to visibility and scrutiny. In this way, and before any functioning automated facial expression analysis system has been invented, FACS transforms faces and their expressions into a field of data, into objects for data processing, with the aim of using the data to make sense of the mysterious inner emotional lives of individuals. FACS informationalizes the face and its expressive movements, laying a grid over its surface, breaking it down into fragments, and assigning each part an arbitrary number and a basic descriptive term that can then be offered up for the construction of new meanings in the psychology of the face and emotion. While the contention is that FACS represents the material dimensions of facial expression, it actually provides a new language of facial analysis and interpretation, one that makes a particular claim to truth based on its instantiation in the form of standardized code.

The *automation* of FACS is significant in that it promises to speed up the expression coding process, standardize classification, make it more precise, and increase the quantity of studies, leading to more data for analysis. Automation has the potential to enable the more widespread deployment of facial action coding practices, extending facial expression processing and its associated behavioral evaluations to a larger number of individuals, thereby generating more data from which to formulate theories for differentiating normal from pathological expressions of affect—that is, for delineating human emotional experience and thereby defining human subjectivity. It also promises to make more readily visible the virtually imperceptible details of facial action that FACS itself attempts to make visible through manual coding. AFEA would ostensibly compensate for the fallibility of human subjective perception, providing a more precise, accurate, objective, scientific analysis of facial affect that in turn could be used to reach more people and a wider range of settings. But just as facial recognition systems help to construct the identities they claim to merely represent, automated FACS systems, if made to function effectively, would likewise push back out into the world and onto faces a uniform system for classifying facial expressions, creating a standardized framework for determining their meanings and what they reveal about the affective lives of human beings.

## *The Mobilization of Affect*

In order to understand the full significance of AFEA as a technology for measuring, defining, and structuring human affect, it is necessary to consider some of its envisioned uses. It is not the technical processes of classification and automation alone that define the meaning and effects of the technology. Equally if not more important is the application of a new, computational technique of emotion measurement to address a set of perceived problems in the domain of individual psychology. Like other forms of psychological measurement, AFEA is envisioned as applicable to a range of institutional uses beyond basic research in psychology.[97] What Nikolas Rose calls "the *techne* of psychology"—including psychological experts, vocabularies, evaluations, and techniques—is now fully integrated into the spheres of the workplace, marketplace, electoral process, family life and sexuality, pedagogy and child rearing, law and punishment, and the medico-welfare complex.[98] Considering the deployment of the *techne* of psychology in this wide range of institutional settings, Rose calls attention to the intrinsic relationship between the emergence of the "psy knowledges" and the history of liberal forms of government, understood not in the narrow sense of state institutions but in the broader sense of the government of conduct.[99] If the rise of "the psychological" is "a phenomenon of considerable importance in attempting to understand the forms of life we inhabit at the close of the twentieth century," as

Rose maintains, then the potential uses of automated facial expression analysis in this expanding psychological domain deserve special attention.[100]

More to the point, the interest of psychologists, computer scientists, and other social actors in the development of AFEA needs to be understood in relation to a wider set of emerging strategies and techniques for the social regulation and mobilization of affect. According Nigel Thrift, late capitalist societies are seeing an expansion of the political sphere to include a new level of emphasis on the affective domain, including the formation of biopolitical strategies that target the embodied dimensions of emotion.[101] While affect has long been "a key element of politics and the subject of numerous powerful political technologies," Thrift argues, evidence of the rise of a particular late capitalist brand of "affective politics" can be found in a number of developments.[102] Most basically, the form of politics itself has changed, with a redefinition in what counts as political to include an intensified focus on the expression of affect and affective appeals.[103] We have also seen the pronounced "mediatization of politics," or the rise of a media-saturated political sphere where political presentation conforms to mediated norms that emphasize key affective sites like the face and the voice, and enlist the performance of emotion as an index of credibility.[104] Moreover, systematic knowledges for the creation and mobilization of affect have become an integral part of everyday life in developed societies. Thrift emphasizes the careful design of urban environments, but similar affect-oriented design strategies are also employed in malls, nightclubs, movie theaters, media production, domestic spaces, and a wide range of other settings. While there is a long history behind these efforts to engineer affect in cities, architectural design, and media, what is unique today is "the sheer weight of the gathering together of formal knowledges of affective response," and the wide spectrum of available technologies for measuring and managing affective behaviors.[105] There is of course no way of guaranteeing particular affective responses in any of these contexts, but "the fact is that this is no longer a random process either."[106]

As a technology of measuring and differentiating the dimensions of affect as they manifest in facial expressions, AFEA is especially suited to what Thrift calls "microbiopolitics," the extension of biopower to a newly detailed level of analysis and intervention into "bare life."[107] The desire to visualize and analyze bodily movements and intensities at the micro-level motivated the photographic work of Muybridge and Marey in the late nineteenth century, as well as early scientific studies of facial expressions. Today, new digital technologies are being designed to bring into existence even more infinitesimal levels of embodiment, making visible and knowable small spaces and times that were previously imperceptible and beyond the threshold of knowledge and power. With the invention of new technologies for measuring the micro-physiological dimensions of affect, we are seeing "the growth of new forms of calculation in sensory registers that would not have previously been deemed 'political.'"[108]

One place where we find the biopolitical potential of AFEA is in its envisioned applications for advancing the psychological science of facial expressions—in the basic and applied research on faces in the discipline of psychology. Although psychologists studying the face may have a narrow range of interests and motivations, the social uses of their work extend beyond the research setting. Since their inception, the psychological sciences have provided "the devices by which human capacities and mental processes could be turned into information about which calculations could be made," making possible "the rational regulation of individuality."[109]

As we have seen, the interests of psychologists studying facial expressions have been a defining feature of AFEA research and development. For psychologists, the automation of the Facial Action Coding System promises to facilitate the production of a wealth of new knowledge about facial expressions and their relationship to emotion and other dimensions of individual psychology. Like other methods in the psychological sciences, FACS itself is a technical device for "exercising a certain *diagnostic* gaze, grounded in a claim to *truth*, asserting technical *efficacy*, and avowing *humane* ethical virtues."[110] FACS methodology facilitates the production of what Rose calls "calculable individuals," making facial affect a technical object defined through a computational process that can precisely mark out individual differences in psychological makeup. The aim is to capture and differentiate facial expressions, to open them up to psychological examination so that individuals themselves can likewise be opened up, their interior selves made visible and amenable to analysis, classification, and intervention. The essential truth of the individual is ostensibly laid bare through the computational analysis of the face. FACS aims to make the interior of the individual more visible, decipherable, and knowable, adding to the knowledge of individual subjectivity, how the self works, and how the individual can be made more adaptable to therapeutic techniques.

Consider, for example, a study that employed manual FACS methods in order to examine "facial expressions of emotion and psychopathology in adolescent boys."[111] Here psychologists sought to determine whether a sample of school boys "with externalizing problems" expressed more anger than a "nondisordered" group, or a third group identified as having a tendency to "internalize" their emotional problems.[112] The boys were placed into categories on the basis of their teachers' assessments of them using the Achenbach Child Behavior Checklist.[113] Findings of this study not surprisingly showed that "adolescents who were reported by their teachers to be more prone to delinquent and aggressive behavior expressed a higher ratio of facial expressions of anger that were of greater magnitude."[114] If a functioning AFEA system were applied in this research, it would facilitate the faster evaluation of a larger number of children and the production of more data for defining the range of normal and abnormal levels of anger expression. It would also ostensibly provide a more precise and objective measurement of boys' expressions of anger, a means of making automatically perceptible the minute motions and intensities of their faces, offering up these newly visible details of embodiment as a measure of their internal emotional lives. It is not difficult to conceive of AFEA serving as a diagnostic tool to classify children according to whether they display normal or pathological levels of anger, and those automatic assessments in turn forming the basis for determining what sort of intervention is necessary in order to normalize their behavior. By automatically processing facial expressions in order to determine types and intensities of psychopathology, AFEA could conceivably perform an "auto-therapeutic" function akin to the work that some psychiatrists thought Weizenbaum's language analyzer might be able to perform in the 1960s—in other words, some of the intellectual labor of psychotherapy. AFEA would in this way provide a technical means of imprinting the social regulation of affect onto children, not simply measuring how they actually feel and express themselves, but using the computational analysis of their faces to define and ultimately regulate their appropriate range of emotional responses to the world.

Although there may in fact be good reasons to monitor and regulate anger expression in adolescent boys, it is nevertheless necessary to understand how the machinery of expression analysis is constructed, how this seemingly technical process is in fact bound up with social conventions that define appropriate levels of anger and other forms of emotional expression, and especially how the social regulation of affect is tied to the exercise of political power.[115] Envisioned applications of AFEA would be especially useful techniques to facilitate what Toby Miller calls the "psy-function," a shifting field of knowledge and power over the mind that is now a defining feature of contemporary U.S. society.[116] The "psy-function" comprises psychoanalysis, psychology, psychotherapy, psychiatry, social psychology, criminology, and, of particular concern to Miller, psychopharmacology. Miller examines the connections between this "psy-function" and the rise of neoliberalism. Like other social theorists discussed in previous chapters, Miller argues that the

United States is essentially experiencing a second Gilded Age, with wealth systematically distributed upward and risk off-loaded onto the working, middle, and lower classes. To cope with their declining social and economic security, people are invited to perpetually reinvent themselves, but in circumscribed and problematic ways—not in terms of more active, organized political citizenship but through avenues like psychopharmaceuticals, religion, and consumption. Children have not escaped these forms of human reinvention, and in fact they are increasingly subjected to pharmaceutical and other interventions designed to address what appear to be a growing number of childhood anxieties and "psychopathologies." The computational analysis of facial expressions has a particular role to play in generating these psychologically oriented models of self-refashioning. The FACS-based psychological study of the face makes its own contribution to the psy-function, one expression of the effort to harness the psychological analysis of emotion to the production of models for self-reinvention.

Functioning AFEA systems would be especially useful for the affect-centered study of consumption. Psychological techniques for measuring emotions have long been applied to the field of consumer behavior, used to analyze affective responses to commercial messages in order to design more persuasive strategies and purchasing inducements. Emotion is widely recognized in marketing as a key component of consumption, with the attachment of emotions to products and services for sale in the marketplace as the goal of most advertising. In *The Marketing Power of Emotions*, John O'Shaughnessy and Nicholas Jackson O'Shaughnessy explain the prevailing view in the field that strong brand loyalties depend on the emotional connections consumers establish with brands, and that effective marketing techniques require a better understanding of how consumer choices are guided by emotions.[117] In the language of market research, "NERS" scores represent "net emotional response strengths" and are estimated by drawing on a variety of emotion measurement methods, including skin conductivity tests, heart rate measurement, gaze tracking, brain scans, and of course the analysis of facial expressions.[118] Although the labor-intensity of FACS coding makes it prohibitive for large-scale market research studies, functioning AFEA systems would allow for more widespread use of facial expression measurement techniques to assess the affective responses of consumers to advertising and other marketing strategies.

But the use of AFEA for market research purposes is not simply about measuring emotional responses to products or encouraging brand loyalty. Specific studies that measure physiological manifestations of emotion in response to mediated commercial messages certainly have these narrow goals. In a broader sense, however, the development and application of a wide range of affect measurement technologies for analyzing in increasingly minute detail the way people respond at the embodied level to variations in marketing strategies has the more consequential goal of channeling human affect to commodity consumption more generally. It is not any particular emotional response to a product or advertisement that is most significant, but the more cumulative and broad-based effort to harness affect to the purpose of advancing the project of consumer capitalism, making and remaking people in the mold of the emotional consumer. The effort to engineer affective responses in marketing, advertising, and other commercial media provides a scientific foundation for U.S. makeover culture, which offers a model of self-invention through consumerism; "the deepest recesses of the human psyche" are treated as "means for the expansion of the commodity universe."[119] Such marketing strategies are biopolitical to the extent that they aim to bind embodied affective experience to the emotional labor of consumption.

Another important site where we find the biopolitical potential of AFEA is in its envisioned applications for deception detection. According to Ekman, he and his colleagues developed FACS expressly for this purpose. Ekman suggests that the main impetus behind his eight-year-long project to develop FACS was to make advances in deception research. In his trade book *Telling Lies: Clues to Deceit in the Marketplace, Politics, and Marriage,* he explains that, after conducting a study in which subjects were unable to accurately determine whether student nurses portrayed in videos where lying or telling the truth, he and his colleague Wallace Friesen decided they needed to come up with a more precise system for capturing "the fleeting hints" of "concealed emotions," or the "facial leakage" that most people seem to overlook.[120] They set out to develop a "comprehensive, objective way to measure all facial expressions," according to Ekman, because they knew that "uncovering facial signs of deceit would require precise measurement."[121] The manual version of FACS is already used not only for deception research in psychology but also for deception detection in law enforcement and security settings. For example, "Behavior Detection Officers" working for the U.S. Transportation Security Administration's SPOT program (Screening Passengers by Observation Techniques) are trained in FACS and use it along with other methods in order to identify bodily signs of deception and abnormal affective behaviors among people passing through airport security.

The successful automation of FACS is viewed by these institutional users as a promising prospect for enabling its more widespread diffusion, facilitating efforts of law enforcement, security agencies, businesses, and other social actors to detect deception in a host of settings where trained human coders are in short or no supply. In an October 2006 article published in the *Washington Post*, Ekman extolled the benefits of the SPOT program and predicted the imminent transition to automated forms of deception and anomaly detection: "Within the next year or two, maybe sooner, it will be possible to program surveillance cameras hooked to computers that spit out FACS data to identify anyone whose facial expressions are different from the previous two dozen people."[122] As Ekman makes clear, the automation of FACS for deception detection—in this case based on apparent anomalies in facial expression among groups of individuals—is not just an envisioned possibility but also an explicit policy priority. (The research groups at UCSD and CMU working on the automation of FACS have received funding from the CIA as well as the NSF.[123])

In addition to allowing the more widespread diffusion of FACS for deception detection, the automation of FACS is seen as especially useful for identifying facial actions that are virtually imperceptible to most human observers. While FACS can yield many secrets about a subject's true feelings that he or she might be trying to hide, according to Ekman, the "most tantalizing source of facial leakage" is the "micro-expression," an almost undetectable facial action that "flashes on and off the face in less than one-quarter of a second."[124] While these suppressed expressions are virtually impossible for untrained observers to detect, AFEA systems potentially could be programmed to capture and identify them, replacing the need for highly trained human observers and painstaking analysis of slow-motion video. According to one researcher, "An automatic facial expression system could scan large quantities of film for micro expressions in a relatively short period of time."[125] Video from a wide range of sources could be processed for detection of micro-expressions, from interrogations to employment interviews and courtroom testimony, as well as video of public figures giving speeches or being interviewed by the news media. One technical article on automatic facial expression interpretation suggests that AFEA could allow real-time lie detection systems to operate "in court-rooms, police headquarters, or anywhere truthfulness is of crucial importance."[126]

Here, what Thrift calls the "mediatization" of these public spaces not only makes them conform to mediated norms of affect performance, it also makes the bodies represented in these recorded spaces and times available for the computational analysis of their affective credibility. AFEA provides a means for the automated, computational processing of volumes of video in search of virtually imperceptible micro-expressions, aiming to lay bare the most minute traces of deception. The measurement of the alleged physiological bases of deception takes on an especially biopolitical cast, as deception becomes a mark of both behavioral and biological deviance, and even a threat to the nation, as it became in the post-9/11 environment.[127] Of course, deception detection has long been a deeply contested area of research and practice, and this disputed terrain is instructive for understanding both the promise of AFEA and its fundamental problems, especially when accompanied by claims to its computational objectivity.

The polygraph test is the paradigmatic lie detection technique, and a protracted struggle has played out historically over its scientific and legal legitimacy. In reality, no credible scientific evidence exists to support the polygraph's ability to determine whether someone is lying or telling the truth. There is no proof that the polygraph measures anything other than a set of physiological functions, and the general inadmissibility of polygraph test results in U.S. court proceedings stems from this lack of scientific legitimacy. The decision in *Frye v. United States* (1923), a precedent-setting case for the admissibility of scientific evidence, excluded lie detector test results based on their lack of "standing and scientific recognition among physiological and psychological authorities."[128] More recent evaluations likewise have found that "it is not possible to use the existing literature to provide a satisfactory error rate estimate" for polygraph tests,[129] although according to one review of actual tests conducted in the field, "the innocent were called truthful only 53 percent of the time."[130]

Still, the polygraph's lack of scientific and legal standing has not prevented its adoption in a range of institutional settings, especially in employment, law enforcement, and military interrogations. Nor has it completely undermined its ability to generate productive effects, especially when individuals subjected to the test are convinced of its truth-telling (or lie-revealing) capacity, and when those who administer the test make decisions based on the derived results. The historian Ken Adler has traced the history of the polygraph in the United States, showing how the effort to persuade people of the machine's ability "was itself a prerequisite for the machine's success. As its proponents acknowledged, the lie detector would not distinguish honest Americans from deceptive ones unless those same Americans believed the instrument might catch them."[131] The polygraph is a technology of highly questionable validity, yet one that continues to be put to productive use.

Proponents of newer deception detection technologies, like brain scans or functional magnetic resonance imaging (fMRI), claim that these technologies offer more accurate measures of deception than the polygraph. But new techniques of deception detection are unlikely to resolve the fundamental problems associated with scientific methods that target the body for clues about the state of mind of the individual. Although proponents of fMRI claim that it is a much more advanced technology of deception detection, offering direct access to the workings of the brain, in fact the technology functions much like the polygraph in that it measures physiological processes as indirect indicators of deception.[132] "Despite rhetoric to the contrary," writes Melissa Littlefield, "brain-based detection is as equally invested in correlating deception and truth with the body as was traditional polygraphy."[133] In the case of AFEA, facial expressions are translated into coded information for data processing, a process that ostensibly provides a more detailed, precise, and objective measurement of expression. But much like fMRI, FACS and AFEA "measure" deception indirectly rather than offering direct access to an individual's thoughts and feelings.

Of course, this will not prevent FACS and AFEA from being used to make authoritative claims about honesty and deception, any more than similar problems prevented the use of the polygraph for this purpose. Like the polygraph, functioning AFEA systems depend, now and in the future, on a belief in the capacity of the technology to do what it purports to do, regardless of whether it can do so with complete "accuracy."[134] The claim to accuracy rests on both the belief in the computational objectivity of FACS and on the *biologization* of honesty and deception: defining these culturally variable and historically changing practices of the self, which take shape in humans' interaction with one another, as physiological processes that can be located and coded in individual bodies.[135] The biologization of affect in general, and honesty and deception in particular, transforms these social and cultural phenomena into measurable objects of science, thereby enabling them to be incorporated into the domain of biopolitics. In its envisioned application for deception detection, AFEA promises to assist with the government of individual subjectivity in the name of truth, functioning quite literally as a "technology of truth" that employs a computational model of facial expression analysis to make determinations about truthful versus deceitful conduct in a wide range of settings—potentially anywhere where faces are captured in images.

## *Computing Faces and Emotions*

Although they overlap, the aims of computer scientists working on AFEA research and development are not precisely the same as psychologists studying facial expressions, marketing specialists, or the various social actors invested in its applications for deception detection. Some computer scientists are interested in their own basic research, while others are interested in applying AFEA to develop more sophisticated forms of computer vision and human-computer interaction (HCI). Computer scientists envision AFEA and other affect-sensing technologies as integral components in the design of immersive computing environments where computers respond and automatically adjust to the verbal and physiological signals of human beings. One of the major trends in computer science discussions about HCI is an emphasis on "human-centered computing"—the design of computer systems "that can unobtrusively perceive and understand human behavior in unstructured environments and respond appropriately."[136] The goal of "human-centered computing" is "to design environments which do not impose themselves as *computer environments*, but have a much more natural feeling to them."[137] "The main motivating principle," according to researchers in this area, "is that computers should adapt to people rather than vice versa."[138] Where computer-centered HCI requires human beings to adapt their behaviors to computers, so-called human-centered computing applications aim to allow humans to interact naturally with invisible computer programs through transparent interfaces. "The time when rooms, walls, desks, and blackboards will be invisibly given computing power" apparently is not far off.[139] The computer's ability to "see" and respond to human affective states is viewed as central to the aim of human-centered computing and immersive computing environments, and functioning AFEA systems represent one technique in the broader effort to achieve these aims.[140]

Although the priorities that shape the application of AFEA to the development of more sophisticated and human-friendly forms of human-computer interaction differ in many ways from those that shape its applications in psychology, market research, and deception detection, these different potential uses of the technology nevertheless share some important characteristics. So-called human-centered approaches to system design may in fact be more attuned to certain human needs and behaviors than conventional forms of computing. But much like the human-centered approach to information retrieval discussed in chapter 4, they likewise aim to harness particular forms of human subjectivity, technical skill, and mental and emotional labor to the task of building systems that in turn can be used to monitor individuals with increasing ubiquity and precision. And while the claim of human-centered computing is that it provides a more transparent interface, the forms of systematic data collection that accompany this form of "interactivity" become less visible to those individuals who are ostensibly empowered by new human-centered technological configurations. Immersive computing environments exhibit what Jay David Bolter and Richard Grusin call the "double-logic of remediation," aiming "to achieve immediacy by ignoring or denying the presence of the medium and the act of mediation."[141] But while Bolter and Grusin theorize this double-logic as a unique characteristic of new media, in fact this effort to deny the act of mediation can also be understood as an expression of what Langdon Winner has identified as a basic tendency of modernism: the adoption of technical design strategies that "conceal and obfuscate important realms of social complexity."[142] Human-centered computing environments that automatically respond in a "natural" way to users' affective states would render obscure the social complexity of affect, especially insofar as the emotion classification systems and layers of technical integration built into these programs function as automatic and more or less seamless parts of their design.

Despite the passive model of subjectivity implicit in the paradigm of human-centered computing, the discourse of HCI nevertheless promotes an ideal of an active, tech-savvy computer user engaged in playful and productive forms of media "interactivity," sharing much in common with the tech-savvy subject we encountered in chapter 4. The discourse of HCI similarly invokes a mode of subjectivity whereby individuals are expected to continuously experiment with and adopt new technologies as part of their practices of the self. In fact, some HCI designers advocate the use of AFEA as a means of helping users better understand their emotions and individual psyches, suggesting that "affective computing" applications can be used for emotional and expressive self-assessment. According to researchers in the Affective Social Computing Group at Florida International University, affective computing offers more than the practical benefits of invisible computation designed into the built environment: "We might also find an opportunity to better understand ourselves by building multimodal tools with increased awareness of the user's states."[143]

Like the nineteenth-century scientists who analyzed the physiological manifestations of their own emotions, creating an "emotional economy" in the laboratory, computer scientists themselves have been the first to experiment with new affect-sensing technologies. But once again, the implications and potential uses of these self-assessment applications of AFEA extend beyond the laboratory to other social spaces. For example, a proposed "human-human communication trainer" application of AFEA promises to help users make sense of their own facial expressions and develop ways of better communicating with others:

> Very often people are less aware of their own expressive style than their mates and coworkers for the obvious reason that they most of the time do not see their own facial expressions. By providing facial images with their interpretation during a communicative exchange (say over the Internet), the trainer system would enable people to become aware of their own expressions, possibly learning from them, and having the possibility of adjusting their expressive style, or disambiguating their expressions to the people they communicate with.[144]

Here AFEA is posited as a means of enabling individuals to regulate, manage, and manipulate their own facial expressions and "expressive style," functioning as a technology of communicative self-evaluation and self-improvement. Individuals would ostensibly use the AFEA "communication trainer" to train themselves to use their faces to express in more useful, interpretable, and productive ways, while avoiding ambiguous, misleading, or socially inappropriate forms and intensities of expression.

Affect-centered HCI applications have advocates not only in computer science but also in science and technology studies. Kirsten Boehner and her colleagues, for example, find much of value in the forms of emotional self-examination and self-awareness that affect-centered approaches to HCI could afford.[145] Although they critique the conceptualization of affect as information to be processed through individualizing physiological measurement techniques, they nevertheless "welcome the turn to affect in HCI," recommending an "interactional" rather than an "informational" approach to conceptualizing affect in emotion-centered HCI applications. This more interactional and flexible approach to HCI design treats affect not as a biological process to be measured precisely and scientifically, but as a cultural and "intersubjective phenomenon, arising in encounters between individuals or between people and society, an aspect of the socially organized lifeworld we both inhabit and reproduce."[146] This shift in the conceptualization of affect from information to interaction, they argue, would "substantially change the playing field for system design. Instead of designing systems to decode and encode affective signals, systems are set up to engage users in interactions that help them understand, reflect on, and experience their emotions in new ways."[147]

Boehner and her colleagues are right to argue that the treatment of affect as information threatens to miss much of the subtlety and complexity of human affective relations, and that affect is better understood as a social and cultural phenomenon that emerges in humans' interactions with one another and with the world around them. But their interest in the development of computer technologies that would help people better understand their emotions and experience them in new ways resonates with, rather than departs radically from, the application of a computational model to the social regulation and political mobilization of affect. Rather than taking as given the value of emotional self-examination and self-knowledge, it is important to consider why individuals are so persistently encouraged or enjoined to engage in practices of self-assessment. Like other projects that make demands of individuals to know themselves, to take care of themselves, and to better themselves, the drive to develop HCI as a technology of emotional self-awareness and self-improvement suggests that human affect and human subjectivity are envisioned as malleable technological projects, amenable to regulation and reinvention. Insofar as they promise to provide people with some essential knowledge of their interior emotional selves, interactional HCI applications are consistent with, rather than radically different than, other technologies designed to capture, measure, and analyze emotion—technologies that promise to lay bare the affective lives of individuals in order to facilitate the social regulation of affect. The application of affect-sensing technologies to the aim of emotional self-knowledge resonates with a certain regime of the self characteristic of late capitalist societies, one that encourages individuals to continuously examine and reinvent themselves in order to acclimate to their changing social, economic, and technological environment.

This is not to suggest that the social regulation of affect should be viewed as a narrowly repressive project. Forms of affect regulation are an inescapable and productive dimension of modern social relations, inherent in what Elias has called "the civilizing process." But even if we acknowledge the necessity and even desirability of certain forms of affect government, it is nevertheless worthwhile to raise the question of why *computers* would be conceived as a means of helping individuals better understand and manage their emotions. There seems no better way to remake ourselves in the image of computers than to apply them toward our own affective self-understanding, as a means of communicating with ourselves and making sense of our affective experiences. In his analysis of the cultural logic of computation, David Golumbia insists that we must "keep in mind the possibility of de-emphasizing computerization, resisting the intrusion of computational paradigms into every part of the social structure, and resisting too strong a focus on computationalism as the solution to our social problems."[148] If there is an ounce of wisdom in this assertion, and I believe there is, then we might also want to heed this advice when considering computers as solutions to problems of self-knowledge. It seems paradoxical to place so much faith in computers to resolve fundamental problems of communication, especially when the conversation is one we are meant to be having with ourselves, about ourselves.

No matter how well or poorly computers are made to "see" facial expressions, what remains difficult to compute is the issue of empathetic recognition, Weizenbaum's central concern introduced at the outset of this chapter. Although certainly not in every instance, human face-to-face interactions are at least occasionally sites where an ethical regard for others is learned, performed, formulated, and enacted. The philosopher Emmanuel Levinas saw the face-to-face encounter as the primary site of ethics, the moment when we are forced to recognize the other as an autonomous being, distinct from ourselves.[149] David Lyon has likewise appealed for an ethic of "embodied personhood" or "local face-to-face care" as a countermeasure to the impersonal and often inhumane forms of administrative procedure and social ordering inherent in large-scale, bureaucratic surveillance systems.[150] These writers share Weizenbaum's concern for a moral and ethical obligation to other people that seems to arise most acutely when humans are in one another's physical presence.[151] While we should be cautious about nostalgically privileging face-to-face communication as morally superior to mediated forms, there may in fact be unique dimensions to human affective relations that are not entirely amenable to computation. Computation requires facial expressions and their analysis to be defined and constrained in particular ways, and translating into computational form the widely variable human capacity for making facial expressions and interpreting the expressions of others involves incorporating into computational models a considerable amount of expressive subtly and ambiguity—things with which computers have not proven particularly adept. Here the crucial distinction that John Durham Peters makes between "communications" and "communication" is instructive.[152] Drawing on Raymond Williams, Peters defines "communications" as "the institutions and forms in which ideas, information and attitudes are transmitted and received." "Communication," in contrast, is "the project of reconciling self and other."[153] "The mistake," writes Peters, "is to think that communications will solve the problems of communication, that better wiring will eliminate the ghosts."[154]

The problem is not that computer programs will sometimes fail to "accurately" read facial expressions. This would not make them any different than humans. Instead, the problem lies in the preoccupation with precision, accuracy, and objectivity characteristic of computational logic, and the kind of political priorities and worldviews this logic empowers. AFEA translates the subtle, ambiguous, relational, and ethical dimensions of human affective relations into discrete, measurable objects in order to make them more amenable to information processing. This process cannot help but transform the meaning and experience of affective relations, remaking them as technical processes that exhibit a level of precision, differentiation, and standardization uncharacteristic of their complicated, culturally and historically variable, analog forms. To quote again from Katherine Hayles, "a simplification necessitated by engineering considerations becomes an ideology in which a reified concept of information is treated as if it were fully commensurate with the complexities of human thought," or in this case, the complexities of human affect.[155]

In his ponderings on the philosophical implications of his ELIZA language analyzer, Weizenbaum suggested that all technologies more or less transform our affective relations with one another and with ourselves. In order to operate instruments skillfully, he noted, a human being must internalize aspects of those instruments into his or her "kinesthetic and perceptual habits": "In that sense at least, his instruments become literally part of him and modify him, and thus alter the basis of his affective relationship to himself [*sic*]."[156] If every technology that we integrate into our lives contributes to changes in our affective relations with one another and with ourselves, modifying our embodied experiences, this point would seem to doubly apply to technologies specifically designed to capture, measure, and manipulate our affective lives. Bruno Latour uses the term "prescription" to refer to the behavior imposed back on humans by nonhuman delegates.[157] Prescription often takes banal forms, like the way we learn to adapt our manner of walking to a new pair of shoes, or our computer work patterns to particular software glitches that we are unable to resolve. But it can also take more consequential forms, and in fact a series of minor adaptations can ultimately amount to significant changes in the way we live our lives and who we take ourselves to be. This makes it especially important to consider the ways that AFEA could either intentionally be used to prescribe certain affective responses, or could lead inadvertently to changes in human affective relations. Users may learn to adapt their own more complex and subtle forms of facial expression and affective behavior to accommodate computational models, and uses of the technology may become divorced from a sense of the complexity, ambiguity, and ethical dimensions of human affective relations. It is of course also conceivable that AFEA might be used for socially beneficial aims, intentionally or unintentionally prescribing back on human beings positive changes to their affective conduct and providing avenues to new emotional experiences. But the questions immediately follow as to what constitutes a social benefit, what determines the range of appropriate affective experience, and why individuals are compelled to greater self-knowledge and perpetual self-reinvention. When the issue is something as central to human agency and subjectivity as our affective interactions with one another and with the world around us, these questions seem especially crucial to consider.

# Conclusion

> If we consider the face as a territory rather than as a token or even as a sign, with its appearance, disappearance, and reappearance always written-through with traces of history of its forms and uses, we find it one of the places in representation where the past and the present collide most powerfully, and continually exchange a message of crisis.
>
> —John Welchman, "Face(t)s: Notes on Faciality"

I began this book with reference to a surveillance camera image that circulated in the media after September 11, 2001, allegedly depicting two of the hijackers passing through airport security that morning. The faces of the men in the video frame were barely visible, and that, combined with its sheer banality, may explain why it was such a chilling image.[1] Its low, grainy quality lent it an air of credibility as an authentic representation of the real, but the tiny visual blurs of the men's faces made them appear not just unidentifiable but indecipherable. The image seemed to provide only more questions, rather than definitive answers, about the men's true identities and motives. I chose this particular image as a frame for the book because it carried with it the regretful yet hopeful idea that facial recognition technology might have identified the men in the video before they carried out their plans, raising the question of what would have to happen to make such an automated surveillance system functional. While it should be clear now why the claim was problematic on many levels, it nevertheless brought into sharp relief the art historian John Welchman's contention that faces and their representations are territories where past and present constantly collide, and in the process exchange a message of crisis.

In the introduction to their nicely illustrated book *In the Eye of the Beholder: The Science of Face Perception*, Vicki Bruce and Andy Young provide a disclaimer for their use of artistic renderings of faces throughout the text to explain scientific principles of human face perception. They employ these portraits—not just photographs but also paintings, drawings, sculptures, and other facial representations—to make "scientific points" about the visual perception of faces, "and do not consider the social or artistic factors involved in the interpretation of portraits at all."[2] The "point of view of art theory," they insist, "is not one which matters for the points we wish to make."[3] Still, while they see "social and artistic factors" as having little or no bearing on the science of face perception, they do believe that the reverse is true, that science has something to offer a broader understanding of how humans see the face: "We present and discuss scientific findings because we believe these can inform our understanding of what goes on when anyone (artist or non-artist) looks at a face or a realistic portrait, not because we want to constrain what artists do."[4] Bruce and Young may not want to constrain what non-artists do with faces either, but their disclaimer suggests a greater concern with separating science from art than with making a statement about the limits of science to explain how humans perceive the world. In fact, the scientific work they present in their book largely assumes that face perception is based on a universal way of seeing the face. For them, face perception boils down to a set of physiological absolutes that are clearly distinguishable from "social and artistic factors." They are not primarily worried about constraining what nonartists do with faces because ultimately, in their view, everyone perceives faces in essentially the same way.

This view that face perception is essentially a physiological process—separable from the widely variable cultural practices and forms of social organization in which faces and bodies are embedded—has special affinities with the effort to simulate face perception in computers. Much of the computer science literature tethers the proposition that computers can be programmed to "see" the face to the existence of a universal human form of face perception and its fundamental basis in our biological "programming." In addition, just as many experts in human face perception maintain that human evolution has produced this universal, physiological process that can be explained definitively via a unified science, many computer scientists suggest that the move to model face perception in computers is part of the natural progression of science and technology. The opening paragraph of Stan Z. Li and Anil K. Jain's *Handbook of Face Recognition*, for example, offers precisely this sort of explanation:

> Wide availability of powerful and low-cost desktop and embedded computing systems has created an enormous interest in automatic processing of digital images and video in a number of applications, including biometric authentication, surveillance, human-computer interaction, and multi-

media management. Research and development in automatic face recognition follows naturally.[5]

Here the emergence of automated face perception technologies is explained in cursory evolutionary terms, as technologies beget new technologies. It is simply the availability of computing systems that leads to interest in technologies for automatically processing images, since their utility and desirability are self-evident. In this view, recognizing faces and facial expressions are basic human processes, so it is only natural to program computers to do these things, once the technological conditions are ripe.

Two arguments against bothering to write a book like the one in your hands can be found precisely in the two positions just outlined: first, that face perception is a universal, unchanging physiological process separable from historical, social, and cultural influences; and second, that the development of automated face perception technologies simply follows naturally from other technical developments, part of an inevitable, evolving process of computerization. The argument elaborated in this book maintains that the first position is not an adequate explanation of the culturally variable and historically changing ways that human beings use their faces or interpret the faces of others, and the latter position is not a sufficient way of explaining current institutional interest and investment in automated face perception technologies. Although there is of course a physiological dimension to visual perception, there is no universal way of seeing, and no universal way of seeing the face. And while certain technical developments have created both the possibility and areas of need for automated face perception, the conditions of possibility for these technologies cannot be reduced to those technical developments alone.

Instead, automated face perception technologies take their place in a long history of representational practices of the face, and both the roles these technologies perform and the forms they take can only be adequately understood in relationship to that history. It has not been the purpose of this book to provide a complete history of facial representation, but it is nevertheless fundamental to the argument presented here to recognize that the automation of face perception is inextricably tied to a set of historically and culturally contingent political priorities, institutional practices, and social conventions. History and social theory have something to tell us about the automation of face perception. Likewise, any system for representing the face tells us something about the society and historical moment that produces it. Welchman goes so far as to suggest that "the face is probably the primary site of visual representation, and has shaped the very conditions of visuality."[6] Although too sweeping a statement, since no system of visual representation can lay claim to constituting *the very* conditions of visuality, it is worth contemplating this idea, because any force that shapes human visual practices, however contingent or contested, shapes our ways of understanding the world and our places within it. The face has been a special object of attention in the organization of visual practices and the development of visual media technologies, and technologies designed for representing and analyzing the face have played a special role in defining and redefining what it means to be human.

Beyond being deficient accounts, these two scientific and technical positions can serve ideological functions, however inadvertently. Claims to universality can often be imperialistic and totalizing, flattening out differences in ways that structure the range of human possibilities. The systems devised for representing and interpreting faces are systems for classifying and defining categories of humanness, and the effort to automate any such system is almost inevitably an effort to standardize and disseminate it, to apply that system across contexts, making it reach further in space and time in order to classify and define more faces and more people. This is one sense in which, in many of their envisioned applications, automated face perception technologies are not all that different than nineteenth-century physiognomy and other classifying forms of facial analysis that came before. The technological projects outlined in this book articulate the myth of the digital sublime to more antiquated yet enduring cultural assumptions about the relationship between faces and identities, or between faces and the essential qualities of individuals. They also build on, in material ways, existing practices of facial identification and analysis, retrofitting them with computational systems in order to make more definitive claims to truth about facial images and their embodied human counterparts.

Whether intentionally or not, applying a standardized way of seeing the face to others can deny those embodied subjects their independent existence, their own ability to speak for, represent, and exist for themselves. In *Totality and Infinity*, Levinas discusses "totality" in terms of the violence done to others when they are made to fit rational categories and denied their own autonomous position as unique and separate individuals.[7] Levinas sees the face-to-face encounter as the primary site of ethics, the moment when we are forced to recognize and respect the otherness of the Other. The presence, in the flesh, of an entirely separate individual forces us to acknowledge their difference, the inescapable fact that they are like us, but fundamentally different. We cannot fully know another person, and it is when we face another person in an immediate sense, look that person in the face, that this inability to ultimately reconcile self and Other is most acutely manifest. The need to address the problem of "lost presence"—of bodies missing from communicative contexts—is a problem that affects everyone who engages in mediated exchanges, bound up with the more general problem of maintaining forms of social organization and cohesion in societies of increasing scale and complexity. But to paraphrase John Durham Peters, it is a mistake to think that more sophisticated *communications technologies* will eliminate the fundamental problems of *communication*, of humans' relationships with one another, with themselves, and with their worlds.[8]

Another argument against the thoroughgoing social critique of automated face perception technologies offered in this book holds that more advanced forms of surveillance are essential for security and risk management in a world rife with threats to the well-being of civilized society. This position, characteristic of much state security and biometrics discourse, is unapologetic in its politics of inclusion and exclusion, preferring to assume that the world divides into clearly defined categories of belonging rather than entertain questions about why certain groups are deserving of "security" and others not. The former Alaska governor Sarah Palin invoked a variant of this idea in her closing comments at the 2008 vice presidential debate, for example, when she called on U.S. citizens to join her in the fight for their "economic" and their "national security freedoms."[9] Although not three words commonly strung together to form a single phrase, by "national security freedoms" Palin most likely meant the freedom to do what is necessary to ensure the nation's security, the right for the nation to be secure and thus to secure itself from internal and external threats. The phrase invoked a sense of freedom under siege, requiring protection and made possible only by a strong national security apparatus. Intentional or not, it carried a whole series of connotations pointing to the national security strategy adopted in the U.S. "war on terror," from the doctrine of preemption, to waterboarding, to warrantless wiretapping. Articulating the fight for "economic" freedoms to "national security freedoms," Palin's statement also invoked an image of the individual entrepreneurial spirit under attack, in need of safeguarding from the combined threats of taxation and terrorism. She continued:

> It was Ronald Reagan who said that freedom is always just one generation away from extinction. We don't pass it to our children in the blood-

stream; we have to fight for it and protect it, and then hand it to them so that they shall do the same, or we're going to find ourselves spending our sunset years telling our children and our children's children about a time in America, back in the day, when men and women were free.[10]

Palin ended the debate with a threatening vision of the future, where freedom unprotected and un-fought-for becomes a thing of the past. Having allowed the state to grow too large and tax too much and provide too many social services, Americans look back longingly on all the freedoms they have lost—the freedom to start up new businesses and make as much money as they deserve, to go to any doctor they want, to drive any brand of car, on the open road, to any destination. Reagan's fears were similar to Orwell's—the specter of communism haunting the West and threatening to usher in a socialist totalitarian nightmare, crushing the individual and his freedom of choice under the boot of an obese and overly watchful state. From the subject position of right-wing femininity, Palin updated that scary scenario for post-9/11 America, a place and time where it becomes possible to utter the words "national security freedoms" and posit them as something to fight for—indeed, something to kill for.

As David Harvey has noted, the founding figures of neoliberal thought gave primacy to the political ideals of human dignity and individual freedom, and in doing so "they chose wisely, for these are indeed compelling and seductive ideals."[11] The idea that people in the United States and other Western democracies possess more freedom today, and that their freedom is something that was fought for and must continue to be fought for because it is perpetually under attack, is a powerful set of ideas, often exploited for political aims. There are of course ways in which many people living in these societies can be said to have more liberties, and certainly more choices, than people living in other places and times. But as Toby Miller has observed of the U.S. present, the "duality of free choice and disciplinary governance is the grand national paradox."[12] Migrants arrive in the United States "cognizant of the country's extravagant claims to being *laissez-faire*," only to find themselves living in a highly administered society.[13] In the United States and other late capitalist societies, "freedom" has been tethered to the "free market" and that free market has ushered in a massive surveillant assemblage.[14] Whether this increasingly ubiquitous, panoptic sorting apparatus amounts to a "kinder, gentler" version of Big Brother is not entirely clear.[15] The finance industry's demand for more secure forms of identification and access control, the market research industry's demand for more sophisticated consumer tracking techniques, employers' demands for more effective forms of labor control—each of these priorities of the business system is inextricably tied to the more repressive apparatuses of the welfare-prison-immigration-control complex. And the freedom and security of the "innocent majority" is often defined in stark opposition to myriad others who seem to inherently pose a threat to that freedom and security.[16]

The efforts to create automated forms of face perception and to define compelling social needs for these technologies are projects inescapably bound up in the paradoxical relationship between free choice and disciplinary governance. The technologies promise to provide more sophisticated forms of administrative surveillance and access control, sorting out populations according to categories of belonging and risk, and facilitating efforts to determine who is entitled to "free choice" and "security" and who poses a threat to those values. The inherent tensions and contradictions of this strategy of pursuing new forms of surveillance as a means of ensuring security and freedom are falsely reconciled by the assumption that automated face perception technologies naturally evolve out of a process of research and development, and that they are essentially equivalent to, or at least continuous with, existing practices of facial identification and interpretation. In fact, according to the tech-neutrality view, pursuit of these technologies need not be understood as part of a particular security strategy, since they are products of the natural progression of science and technology, part of the inevitable unfolding process of computerization.

The claim that facial recognition technology could have prevented the violence of 9/11 is indicative of the strategy that informs the design of new surveillance technologies: it assumes that the source of terrorism and crime and other social problems can be located in the bodies and identities of specific individuals and groups. A critical point I want to make in closing is that the source of threats to security is not found in the identities or essential qualities of specific individuals or groups—neither in "different" identities nor in difference itself. Instead, it can be found in the political-economic and social conditions that create and reproduce social inequality, and the associated intolerance for difference that so often accompanies those societies—and international contexts—in which risks and rewards are so unequally distributed.[17] It follows that technologies designed to target individual identities, defining them according to standardized classification systems and distributing them according to their location in a hierarchy of value and risk, in all likelihood help to reproduce the threats they aim to address. The most stable, peaceful societies are not societies in which everyone is identifiable all the time everywhere they go. Instead, they are societies in which social and cultural differences are respected, and people have a healthy measure of autonomy and control over their own lives. It may be unrealistic to hope for an equal distribution of wealth, but as many people as possible need to enjoy a decent standard of living, and on balance, hold an optimistic view for their future. As long as these basic conditions of "freedom" and "security" are based on a politics of inclusion and exclusion—as long as they are defined as the exclusive rights of a certain class of individuals over and against others who are classified as threats to that security—then the situation will be fundamentally unstable, insecure, and worst of all, unjust. There is undoubtedly a sense in which we do need to fight for our freedoms, but not freedoms defined strictly to fit the "free market" model, and certainly not our "national security freedoms," which depend fundamentally on negating the freedoms of others.

Finally, I do not wish to conclude by making the determinist argument that the technological projects outlined in this book are leading to the inevitable standardization of facial representation and interpretation; nor do I want to suggest that a totalizing process of facial classification is the nefarious aim of computer scientists or other social actors involved in these projects. As I have emphasized throughout this book, the idea that the widespread deployment of these technologies is an inevitable part of a nonnegotiable future plays an important role in their implementation, but the automation of face perception has never been a foregone conclusion. Despite claims about the inevitability of these technologies, there are formidable obstacles to their institutionalization. There are, in addition, all sorts of unintended consequences and alternative uses that could result from their deployment in various spaces and social contexts. In order to make automated face perception a functioning reality, inventors require the world as their laboratory, and the messy social world of embodied faces can be especially unpredictable and uncooperative. The automation of facial identification cannot definitively stabilize faces or identities; nor is it likely that the automation of a facial expression classification system can standardize and clearly define, once and for all, the complex, changing, and emergent world of human affective relations. A relatively stable institutional framework has taken shape for the development and deployment of these technologies, but it is impossible to predict the full range of human-machine assemblages that might result from their integration into existing practices of facial representation and analysis. For these reasons if none other, surely scientists conducting research in this area need to abandon the argument that face perception is essentially a universal physiological process, and that the development of automated forms is unfolding as part of the natural progression of science and technology.

One last point I wish to make explicitly in closing is that I am not advocating that the projects of automating facial recognition and expression analysis be abandoned entirely, although I would hold that out as one among the range of possibilities.[18] Instead, I would suggest that serious consideration should be given to circumscribing the uses of these technologies as currently envisioned. Neither the imperatives of companies that stand to profit from these technologies, nor those of the institutions that are their primary early adopters, should take precedence over a thoroughgoing legal and ethical debate about their appropriate forms and uses. This debate needs to reach beyond the limited legal framework of privacy rights to include ethical arguments that identify the forms of structural inequality that new surveillance technologies are often designed to reproduce, even as they appear to function in radically individualizing ways. In addition, the design of these technologies needs to be more widely understood at every level of their development. As Lucas Introna has argued, the social costs of implementing facial recognition systems are not well understood because the methods and choices involved in their design are especially opaque.[19] Algorithm, software, and system development involve processes of closure whereby design decisions are black-boxed, or enclosed in complex socio-technical networks that are virtually impossible for nonexperts to understand. This in turn creates opportunities for silent, invisible micro-politics to be built into these systems, in turn instantiating themselves in the infrastructure of everyday life.[20] In order to address this problem of technological enclosure, Introna argues for a "disclosive ethics" that would aim to open up the process of system design and use to ethical scrutiny. This book represents one effort in that direction.

I would also hold out the possibility that, if radically reenvisioned, versions of these technologies could have beneficial social uses. The definition of beneficial is of course inescapably contested, but design strategies and applications that do not attempt to incorporate or disseminate totalizing facial classification systems would not necessarily be subject to the same criticisms outlined here. Perhaps alternative forms of these technologies could be developed that would not aim to embody an accurate, objective, all-seeing mode of facial identification or expression analysis. It is also conceivable that forms of automated face perception could be devised that would not aim to resolve, once and for all, the problem of the relationship between image and reality, or between portrait and identity. Automated face perception technologies can no more definitively bind our identities to our bodies, determine what we are thinking or feeling, or define our essential qualities than any other form of facial representation or analysis, except insofar as these ideological claims take hold.

Not everyone interested in the automation of face perception is invested in using these technologies to make claims to truth about the fixed relationship between faces and identities. In her essay on "Mediated Faces," for example, Judith Donath of the MIT Media Lab explains that the incorporation of faces into mediated environments involves a host of complex design problems, and one should not assume that the ultimate goal of facial mediation "is to recreate reality as faithfully as possible."[21] There are many reasons why facial images might be usefully reproduced or simulated in mediated environments, and not every scenario demands a direct visual correspondence with the embodied faces of participants. According to Donath, sometimes the goal is to create forms of mediation that differ from face-to-face communication, because "the computer makes it possible to go 'beyond being there'—to create environments that have features and abilities beyond what is possible in the ordinary everyday world."[22]

Donath is right to call into question the ultimate aim of faithfully recreating faces and the conditions of face-to-face encounters in mediated environments. But following the likes of David Golumbia, Katherine Hayles, Vincent Mosco, Kevin Robins, Joseph Weizenbaum, and others, this book maintains that we need to continue to be skeptical of the enduring claim that digital technologies hold out radical possibilities for transcending the "ordinary everyday world."[23] Given the constraints that computers can impose on social exchange, it makes more sense to recognize the ways that non-computing environments have "features and abilities beyond what is possible" in computer-mediated communication. There are infinitely more ways that computer-mediated environments fall short of the ordinary everyday world than vice versa, and belief in technological transcendence has a dangerous tendency to leave the problems of the material world intact. According to Nicholas Negroponte, champion of the digital sublime, "Your face is, in effect, your display device, and your computer should be able to read it."[24] But if our faces are our display devices, pixilated streams of data to be processed by computers, they are also inescapably parts of our bodies and our selves, embedded in social relations, spacial locations, and specific cultural and historical contexts. Faces and their representations are never stable, standardized objects. Instead, they are hybrid assemblages of material and meaning, markers of identity and sites of affect to be sure, but always situated somewhere, on their way from somewhere and to somewhere else.[25]

# Notes

## NOTES TO THE INTRODUCTION

1. Alexandra Stikeman, "Recognizing the Enemy," *Technology Review* (December 2001), http://www.technologyreview.com/communications/12699/page2/.

2. "Biometric Identifiers and the Modern Face of Terror: New Technologies in the Global War on Terrorism," Hearing Before the Technology, Terrorism, and Government Information Subcommittee of the Judiciary Committee, U.S. Senate, 107th Cong. (November 14, 2001).

3. Pat Gill, "Technostalgia: Making the Future Past Perfect," *Camera Obscura: A Journal of Feminism, Culture, and Media Studies* 14, nos. 1–2 (1997): 161–179.

4. Susan Douglas, *Inventing American Broadcasting, 1899–1922* (Baltimore: Johns Hopkins University Press, 1987): xv.

5. See, for example, Matthew Turk, "Computer Vision in the Interface," *Communications of the ACM* 47, no. 1 (2004): 61–67.

6. There are no book-length treatments of the automation of facial recognition and expression analysis from a sociological, legal, or critical cultural perspective, but there are a number of articles that examine facial recognition technology, in whole or in part. See Phil Agre, "Your Face Is Not a Bar Code: Arguments Against Automatic Face Recognition in Public Places" (2001), http://polaris.gseis.ucla.edu/pagre/bar-code.html; Stephen D. N. Graham, "Software-Sorted Geographies," *Progress in Human Geography* 29, no. 5 (2005): 562–580; Stephen Graham and David Wood, "Digitizing Surveillance: Categorization, Space, Inequality," *Critical Social Policy* 23, no. 2 (2003): 227–248; Mitchell Gray, "Urban Surveillance and Panopticism: Will We Recognize the Facial Recognition Society?" *Surveillance and Society* 1, no. 3 (2003): 314–330; Lucas D. Introna, "Disclosive Ethics and Information Technology: Disclosing Facial Recognition Systems," *Ethics and Information Technology* 7, no. 2 (2005): 75–86; Lucas D. Introna and David Wood, "Picturing Algorithmic Surveillance: The Politics of Facial Recognition Systems," *Surveillance and Society* 2, nos. 2–3 (2004): 177–198; Lucas D. Introna and Helen Nissenbaum, *Facial Recognition Technology: A Survey of Policy and Implementation Issues*, Center for Catastrophe Preparedness and Response, New York University, nd, http://www.nyu.edu/projects/nissenbaum/papers/facial_recognition_report.pdf; Torin Monahan, "The Surveillance Curriculum: Risk Management and Social Control in the Neoliberal School," in *Surveillance and Security: Technological Politics and Power in Everyday Life*, ed. Torin Monahan (New York: Routledge, 2006): 109–124; Clive Norris, Jade Moran, and Gary Armstrong,

"Algorithmic Surveillance: The Future of Automated Visual Surveillance," in *Surveillance, CCTV, and Social Control*, ed. Norris, Moran, and Armstrong (Aldershot, UK: Ashgate, 1998): 255–275; Elia Zureik and Karen Hindle, "Governance, Security, and Technology: The Case of Biometrics," *Studies in Political Economy* 73 (2004): 113–137. For a book-length treatment of the relationship between physical faces and new technologies of facial representation that includes some discussion of facial recognition technology, see Sandra Kemp, *Future Face: Image, Identity, Innovation* (London: Profile Books, 2004).

7. David Lyon, *Surveillance Society: Monitoring Everyday Life* (Buckingham: Open University Press, 2001).

8. James Rule, *Private Lives and Public Surveillance* (London: Allen Lane, 1973): 29.

9. David Burnham, *The Rise of the Computer State* (New York: Random House, 1983); Oscar Gandy, *The Panoptic Sort: A Political Economy of Personal Information* (Boulder: Westview, 1993); Mark Poster, "Databases as Discourse, or Electronic Interpellations," in *Computers, Surveillance, and Privacy*, ed. David Lyon and Elia Zureik (Minneapolis: University of Minnesota Press, 1996): 175–192. See also Graham and Wood, "Digitizing Surveillance."

10. Lyon, *Surveillance Society*; Kevin D. Haggerty and Richard V. Ericson, "The Surveillant Assemblage," *British Journal of Sociology* 51, no. 4 (2000): 605–622.

11. Empirical, policy-oriented research, such as Charles D. Raab's, is more conscious of the difficulties inherent in processes of convergence. See Raab, "Joined-Up Surveillance: The Challenge to Privacy," in *The Intensification of Surveillance: Crime, Terrorism, and Warfare in the Information Age*, ed. Kirstie Ball and Frank Webster (London: Pluto Press, 2003): 42–61.

12. Statements about the inevitability of widespread use of facial recognition technology and "smart surveillance" pervade the critical literature on surveillance. For example, in the introduction to their edited volume *The Intensification of Surveillance*, Kirstie Ball and Frank Webster note that "video camera records from key locations (airports, monuments, public buildings etc.) and film from closed-circuit television cameras at freeway tollbooths, *will be examined using facial recognition techniques* to pinpoint suspicious people." Ball and Webster, "Introduction," in *The Intensification of Surveillance: Crime, Terrorism, and Warfare in the Information Age*, ed. Ball and Webster (London: Pluto Press, 2003): 3. Emphasis added. According to Stephen Graham, "Public spaces and malls in the UK and USA *are increasingly being sorted through face recognition CCTV*." Graham, "The Software-Sorted City: Rethinking the 'Digital Divide,'" in *The Cybercities Reader*, ed. Graham (New York: Routledge, 2004): 329. Emphasis added.

13. William Bogard, *The Simulation of Surveillance: Hypercontrol in Telematic Societies* (New York: Cambridge University Press, 1996): 4–5.

14. Ibid., 3.

15. Kevin D. Haggerty and Richard V. Ericson, "The New Politics of Surveillance and Visibility" in *The New Politics of Surveillance and Visibility*, ed. Haggerty and Ericson (Toronto: University of Toronto Press, 2006): 13.

16. In a related vein, Irma van der Ploeg examines different rhetorical devices and discursive strategies used to define biometric technologies, arguing that "public debates on a particular technology should be considered as part of the very process of constituting technology." Van der Ploeg, "Biometrics and Privacy: A Note on the Politics of Theorizing Technology," *Information, Communication, and Society* 6, no. 1 (2003): 89. Van der Ploeg

focuses on diverging assessments about the status of biometrics as privacy-protecting or privacy-threatening technologies. She finds that they differ most significantly in their conceptualization of biometrics as malleable "technologies in the making," versus as artifacts with inherent attributes and fixed meanings. Ibid., 99. Conceptualizing the technology as still at a flexible stage of development "leaves some urgently needed space for active involvement in the very design of the biometric technology to be installed." Ibid., 98.

17. Mark Andrejevic has argued that the media of communication have become the media of surveillance, especially new "interactive" media technologies that ostensibly promise democratic participation while delivering more effective forms of monitoring. Mark Andrejevic, *iSpy: Surveillance and Power in the Interactive Era* (Lawrence: University Press of Kansas, 2007).

18. Tom Gunning, "In Your Face: Physiognomy, Photography, and the Gnostic Mission of Early Film," *Modernism/Modernity* 4, no. 1 (1997): 25. On the relationship between the analysis of the face and new photographic technologies, see also Kemp, *Future Face*; and Robert A. Sobieszek, *Ghost in the Shell: Photography and the Human Soul, 1850–2000* (Cambridge: MIT Press).

19. Mike Baker, "FBI Delves into DMV Photos in Search of Fugitives," *ABC News: Technology* (October 12, 2009), http://abcnews.go.com/Technology/wireStory?id=8814738.

20. Ying-Li Tian, Takeo Kanade, and Jeffrey Cohn, "Facial Expression Analysis," in *Handbook of Face Recognition*, ed. Stan Z. Li and Anil K. Jain (New York: Springer, 2004): 251.

21. Martin Jay, "Scopic Regimes of Modernity," in *Vision and Visuality*, ed. Hall Foster (Seattle: Bay Press, 1988): 3–23. Jay borrows the term "scopic regime" from Christian Metz, *The Imaginary Signifier: Psychoanalysis and the Cinema*, trans. Celia Britton et al. (Bloomington: Indiana University Press, 1982): 61. On modern/Western "ways of seeing," see also John Berger, *Ways of Seeing* (London: Penguin, 1972); Jonathan Crary, *Techniques of the Observer: On Vision and Modernity in the Nineteenth Century* (Cambridge: MIT Press, 1992); Martin Jay, *Downcast Eyes: On the Denigration of Vision in Twentieth-Century French Thought* (Berkeley: University of California Press, 1993).

22. Kevin Robins, *Into the Image: Culture and Politics in the Field of Vision* (London: Routledge, 1996).

23. Donna Haraway, "The Persistence of Vision," in *The Visual Culture Reader*, ed. Nicholas Mirzoeff (New York: Routledge, 1998): 681, 682. Emphasis added.

24. Suzannah Biernoff, "Carnal Relations: Embodied Sight in Merleau-Ponty, Roger Bacon, and St Francis," *Journal of Visual Culture* 4, no. 1 (2005): 49.

25. Lorraine Daston and Peter Galison, "The Image of Objectivity," *Representations* 40 (1992): 81–128. My point here is the opposite of William Mitchell's argument that new digital technologies are "relentlessly destabilizing the old photographic orthodoxy" and subverting traditional notions of photographic truth. William J. Mitchell, *The Reconfigured Eye: Visual Truth in the Post-photographic Era* (Cambridge: MIT Press, 2001): 223. In other words, digital technologies are not only being used to disrupt the relationship between image and reality, they are also being used to make newly empowered claims to the truthfulness of visual images. However, I do not wish to suggest that the relationship between photographs and reality was ever unproblematic. For the classic photo-realist position, which I argue against throughout this book, see Roland Barthes, *Camera Lucida:*

*Reflections on Photography*, trans. Richard Howard (New York: Hill and Wang, 1981). For a strong critique of Barthes, see John Tagg, *The Burden of Representation: Essays on Photographies and Histories* (Minneapolis: University of Minnesota Press, 1988).

26. See Paul Virilio, *The Vision Machine*, trans. Julie Rose (Bloomington: Indiana University Press, 1994). See also Virilio, *War and Cinema: The Logistics of Perception*, trans. Patrick Camiller (London: Verso, 1989). Virilio argues of the displacement of human perception by machines, driven by the progressive enlargement of the military field of perception and culminating in a "logistics of perception" that merges weapons and cameras, enabling the dominant powers to keep military targets perpetually in sight on a global scale. He traces the interrelated development of weapons and visualizing technologies: "Hand-to-hand fighting and physical confrontation were superseded by long-range butchery, in which the enemy was more or less invisible save for the flash and glow of his own guns. This explains the urgent need that developed for ever more accurate sighting, ever greater magnification, for *filming the war* and photographically reconstructing the battlefield; above all it explains the newly dominant role of aerial observation in operational planning." Virilio, *War and Cinema*, 70.

27. See, for example, Martha J. Farah, Kevin D. Wilson, Maxwell Drain, and James N. Tanaka, "What Is 'Special' About Face Perception?" *Psychological Review* 105, no. 3 (1998): 482–498. While many of the techniques developed for automated facial recognition bear no resemblance to human brain function, computer scientists nevertheless have looked to research in cognitive psychology and neuroscience for guidance in solving, or at least thinking about, difficult technical problems in programming computers to see and recognize faces. Relevant research in these fields has included work examining whether face recognition is a dedicated process, whether face perception is the result of holistic or feature analysis, and the relative importance of various facial features for human recognition of faces. The use of both global and local features for representing and recognizing faces in certain automated recognition systems has drawn on research suggesting that the human brain operates in such a way when it analyzes a face. Other questions that are considered relevant to development of automated recognition concern how children recognize faces, what role facial expressions play in recognition, the role of race and gender, and our ability to recognize faces in images of varying quality. Rama Chellappa, Charles L. Wilson, and Saad Sirohey, "Human and Machine Recognition of Faces: A Survey," *Proceedings of the IEEE* 83, no. 5 (1995): 710–711. In addition, automated facial recognition has been partly guided by research suggesting that the brain processes identities and expressions using functionally distinct neurological pathways, and that the relative lightness and darkness of areas of the face play a role in our assessment of human face representations. Vicki Bruce, Peter J. B. Hancock, and A. Mike Burton, "Human Face Perception and Identification," in *Face Recognition: From Theory to Applications*, ed. Harry Wechsler et al. (New York: Springer, 1998): 51–72.

28. James V. Haxby, Elizabeth A. Hoffman, and M. Ida Gobbini, "Human Neural Systems for Face Recognition and Social Communication," *Biological Psychiatry* 51, no. 1 (2002): 59.

29. John Durham Peters, *Speaking into the Air: A History of the Idea of Communication* (Chicago: University of Chicago Press, 1999): 140–141.

30. Ibid., 141.

31. Durham Peters argues that "the body in the medium" is the central dilemma of modern communications, and the "intellectual and political heart of mass communica-

tion theory" is "the possibility of interaction without personal or physical contact." Ibid., 225, 224.

32. Craig Robertson, "A Documentary Regime of Verification: The Emergence of the U.S. Passport and the Archival Problematization of Identity," *Cultural Studies* 23, no. 3 (2009): 329–354.

33. See Jane Caplan and John Torpey, eds., *Documenting Individual Identity: The Development of State Practices in the Modern World* (Princeton: Princeton University Press, 2001).

34. Donna Haraway, *Modest_Witness@Second_Millennium.FemaleMan©_Meets_OncoMouse™* (New York: Routledge, 1997): 142.

35. Jane Caplan, "'This or That Particular Person': Protocols of Identification in Nineteenth-Century Europe," in *Documenting Individual Identity: The Development of State Practices in the Modern World*, ed. Jane Caplan and John Torpey (Princeton: Princeton University Press, 2001): 51.

36. Ibid.

37. Ibid.

38. Techniques of individuation long precede computerization. In *Discipline and Punish*, Foucault describes the appearance a new technology of discipline in the eighteenth century, the examination, which employed the power of writing to record the "everyday individuality of everyone"; the examination signaled the appearance of a "new modality of power in which each individual receives as his status his own individuality, the gaps, the 'marks' that characterize him and make him a 'case.'" Michel Foucault, *Discipline and Punish: The Birth of the Prison*, trans. Alan Sheridan (New York: Vintage, 1977): 191–192. This process of individuation was central to emerging, liberal forms of governance in the nineteenth century. It involved "turning real lives into writing," as Foucault elaborated, and eventually also into *images* that could be meticulously examined one by one and accumulated into filing systems, amounting to a new representation of society. See Tagg, *The Burden of Representation.*

39. The founding managerial document on mass customization is B. Joseph Pine II, *Mass Customization: The New Frontier in Business Competition* (Cambridge: Harvard Business School Press, 1992). Deemed the "new frontier of business competition" in 1992, mass customization required new technologies to facilitate the mass individuation of populations. "Customization. . . increases the demand for demographic information to a new level," as Mark Andrejevic has noted. "It creates new markets for the fruits of increasingly intensive and extensive forms of consumer surveillance." Andrejevic, *iSpy*, 13. According to Andrejevic, new "interactive" media technologies, like the Internet, digital video recorders, cell phones, and machine-readable credit and debit cards, while celebrated as inherently democratic and empowering for their users, in reality enable the meticulous collection of data about their users' lives by market research firms. See also Matthew Carlson, "Tapping into TiVo: Digital Video Recorders and the Transition from Schedules to Surveillance in Television," *New Media and Society* 8, no. 1 (2006): 97–115; and Joseph Turow, *Niche Envy: Marketing Discrimination in the Digital Age* (Cambridge: MIT Press, 2006). On the "individualization of labor," see Manuel Castells, *The Information Age: Economy, Society, and Culture*, vol. 3: *End of the Millennium*, (Malden, MA: Blackwell, 1998): 72.

40. Caplan, "This or That Particular Person," 51.

41. Stuart Hall, "Cultural Identity and Diaspora," in *Identity: Community, Culture, Difference*, ed. Jonathan Rutherford (London: Lawrence and Wishart, 1990): 222.

42. Nikolas Rose uses the term "securitization of identity" to describe the proliferation of sites where individuals are made responsible for establishing their official identity as a condition of access to the rights and responsibilities of citizenship. Rose, *Powers of Freedom: Reframing Political Thought* (New York: Cambridge University Press, 1999): 240.

43. Matthew Turk and Alex Pentland, "Eigenfaces for Recognition." *Journal of Cognitive Neuroscience* 3, no. 1 (1991): 71.

44. For a brief and accessible explanation of facial recognition algorithms, see Introna and Nissenbaum, *Facial Recognition Technology*, 16–17.

45. Wen-Yi Zhao, Rama Chellappa, P. Jonathan Phillips, and Azriel Rosenfeld, "Face Recognition: A Literature Survey," *ACM Computer Surveys* 35, no. 4 (2003): 400.

46. Introna and Nissenbaum, *Facial Recognition Technology*, 21.

47. Conflating facial signs of pathology with racial and ethnic differences continues today, for example, in the practice of screening photographs of children's faces for signs of Fetal Alcohol Syndrome. See Lisa Cartwright, "Photographs of Waiting Children: The Transnational Adoption Market," *Social Text 74* 21, no. 1 (2003): 83–108.

48. Sobieszek, *Ghost in the Shell*, 126.

49. There are applications of facial recognition that are not database driven, that instead involve matching a face in an existing, stand-alone image to an image taken of a live person. For example, local feature analysis could be used to compare a photo ID to the person in possession of the ID, in order to confirm that the live person is the same individual whose picture is embedded in the document, without recourse to a database search.

50. Allan Sekula, "The Body and the Archive," *October* 39 (Winter 1986): 16.

51. Gunning, "In Your Face," 16. See also Sekula, "The Body and the Archive."

52. On biopolitics and state racism, see Michel Foucault, *The History of Sexuality: An Introduction*, trans. Robert Hurley (New York: Vintage, 1978): 135–159. See also Foucault, *"Society Must Be Defended": Lectures at the Collège de France, 1975–76*, trans. David Macey (New York: Picador, 2003): 239–263.

53. Tom Gunning, "In Your Face," 24.

54. The term "blank somatic surface" is from Sobieszek, *Ghost in the Shell*, 86.

55. In 2008, Ekman's work on deception detection became the inspiration for a new Fox television show, called *Lie to Me*, about a fictional team of lie detection experts led by Dr. Cal Lightman, a character whose skill at catching liars is without fault. The show does powerful work for "the imaginary of surveillant control," suggesting that the techniques for detecting deception developed in the field of psychology provide an absolute, accurate, definitive means of determining what people are thinking and feeling. Lightman always recognizes deception when he sees it on the body and is absolutely never wrong. In real life, Ekman is more careful in his writings, where he makes judicious use of qualifications and cautions against making too facile a connection between facial expressions and lying. It can be very difficult to distinguish between deceit and nervousness, for example, and the behavior of truthful people who are under suspicion of lying is often mistaken for evidence of deception. Paul Ekman, *Telling Lies: Clues to Deceit in the Marketplace, Politics, and Marriage* (New York: Norton): 333.

56. Ian Hacking, "Making Up People," in *Reconstructing Individualism: Autonomy, Individuality, and the Self in Western Thought*, ed. Thomas C. Heller and Morton Sosna (Palo Alto: Stanford University Press, 1986): 222–236.

57. Allan Sekula, "The Body and the Archive," 10.

58. Gilles Deleuze and Félix Guattari, "Year Zero: Faciality," in *A Thousand Plateaus: Capitalism and Schizophrenia*, trans. Brian Massumi (Minneapolis: University of Minnesota Press, 1987): 172.

59. Ibid., 167.

60. Ibid., 167–191.

61. Gilles Deleuze, *Cinema 1: The Movement-Image*, trans. Hugh Tomlinson and Barbara Habberjam (Minneapolis: University of Minnesota Press, 1986): 99.

62. Donald MacKenzie, *Inventing Accuracy: A Historical Sociology of Nuclear Missile Guidance* (Cambridge: MIT Press, 1993): 167–168, 386–387.

63. Ibid., 168.

## NOTES TO CHAPTER 1

1. Takeo Kanade, *Computer Recognition of Human Faces* (Basel: Birkhauser, 1977): 33–34.

2. According to Eric Hobsbawm, the years 1966–1970 were peak years of Japanese economic growth, fueled in part by aid from the United States, which had been "building up Japan as the industrial base for the Korean War and [later] the Vietnam War." Hobsbawm, *The Age of Extremes: A History of the World, 1914–1991* (New York: Vintage, 1994): 276.

3. J. P. Telotte, *A Distant Technology: Science Fiction Film and the Machine Age* (Hanover: Wesleyan University Press, 1999): 164.

4. Langdon Winner, "Who Will We Be in Cyberspace?" *Information Society* 12, no. 1 (1996): 67.

5. Kanade, *Computer Recognition of Human Faces*, 34.

6. Arthur L. Norberg and Judy E. O'Neill, *Transforming Computer Technology: Information Processing for the Pentagon, 1962–1986* (Baltimore: Johns Hopkins University Press, 1996): 255.

7. For a brief history of robotics and computer vision, see ibid., 239–254.

8. On the role of the U.S. military in the development of computers, see Manuel De Landa, *War in the Age of Intelligent Machines* (New York: Zone Books, 1991); Paul N. Edwards, *The Closed World: Computers and the Politics of Discourse in Cold War America* (Cambridge: MIT Press, 1996); Norberg and O'Neill, *Transforming Computer Technology.*

9. "Compensations for lost presences" is John Durham Peters's phrase. *Speaking into the Air*, 214.

10. Started in 1899 as a telephone company, NEC is now an IT company with a global reach, with headquarters in more than forty countries. The largest computer manufacturer in Japan, the corporation and its 334 subsidiaries are also in the business of broadband networking, semiconductors, supercomputing, digital broadcasting, digital signage, home entertainment, and household appliances. See http://www.nec.com/.

11. Dan Schiller, *Digital Capitalism: Networking the Global Market System* (Cambridge: MIT Press, 2000); Schiller, *How to Think About Information* (Urbana: University of Illinois Press, 2007).

12. Nikolas Rose uses the term "advanced liberalism" and reserves a more narrow definition for "neoliberalism." See *Powers of Freedom*, 139–140.

13. A number of scholars have noted the connection between neoliberalism and new information technologies. See, for example, David Harvey, *A Brief History of Neoliberalism* (New York: Oxford University Press, 2005): 3. In "Software-Sorted Geographies," Stephen Graham argues that "the overwhelming bulk of software-sorting applications is closely associated with broader transformations from Keynesian to neoliberal service regimes," 562. See also Schiller, *Digital Capitalism*.

14. Rose, *Powers of Freedom*, 240.

15. Norberg and O'Neill, *Transforming Computer Technology*, 247–251.

16. Virilio, *War and Cinema*, 1.

17. Kanade, *Computer Recognition of Human Faces*, 1.

18. De Landa, *War in the Age of Intelligent Machines*, 193.

19. Very little of Bledsoe's work on computer-assisted facial recognition was published because the funding was provided by an unnamed intelligence agency that did not allow publicity. Michael Ballantyne, Robert S. Boyer, and Larry Hines, "Woody Bledsoe: His Life and Legacy," *AI Magazine* 17, no. 1 (1996): 7–20.

20. Ibid.

21. Kanade, *Computer Recognition of Human Faces*, 4.

22. Bledsoe, in Ballantyne et al., "Woody Bledsoe," 10–11. In 1966, Bledsoe left Panoramic Research to return to academia as a professor of mathematics at the University of Texas, Austin. His work on machine recognition of faces was continued at Stanford by Peter Hart. Ibid., 9.

23. T. Sakai, M. Nagoa, and S. Fukibayashi, "Line Extraction and Pattern Recognition in a Photograph," *Pattern Recognition* 1, no. 3 (1969): 233–248.

24. Chellappa et al., "Human and Machine Recognition of Faces." The dissertation project at Stanford was conducted by M. D. Kelly.

25. Kanade, *Computer Recognition of Human Faces*, 9.

26. Kanade's adviser, Professor T. Sakai, also conducted research in picture processing of human face images. It is Sakai's work that was demonstrated in the "Computer Physiognomy" exhibit at the Osaka Expo '70.

27. A. J. Goldstein, L. D. Harmon, and A. B. Lesk, "Man-Machine Interaction in Human-Face Identification," *Bell System Technical Journal* 51, no. 2 (1972): 399–427.

28. Ibid., 400.

29. According to surveys of the literature, research on facial recognition technology was dormant during the 1980s. In my personal communication with Professor Takeo Kanade, he suggested that this is not precisely correct. I would like to thank Professor Kanade for providing the explanation of what transpired in computer facial recognition research from mid 1970s through the 1980s, which I recount in this paragraph. Any errors in my recounting of his explanation are mine rather than his.

30. De Landa, *War in the Age of Intelligent Machines*, 193.

31. Ibid.

32. Max Weber, *The Theory of Social and Economic Organization*, trans. A. M. Henderson and Talcott Parsons (Glencoe, IL: Free Press, 1947). For a Weberian analysis of bureaucratic surveillance forms, see also Christopher Dandeker, *Surveillance, Power, and Modernity: Bureaucracy and Discipline from 1700 to the Present Day* (Cambridge, UK: Polity Press, 1990).

33. An early hint that photographs might circulate widely as objects and documents of identity beyond the realm of criminal identification came with the popularity in the 1850s and 1860s of the *carte de visite*, a small albumen print of an individual that was exchanged as a novelty or business card.

34. Robertson, "Documentary Regime of Verification." Jane Caplan similarly explains that a "standardized system of identification and recognition . . . is distinguished . . . both by the stability or replicability of its operations and by their relative invisibility, in contrast with older more public regimes that were sustained by the distinctively clothed or marked body and by individual face-to-face encounters." Caplan succinctly describes the function of the identity document: "It is the portable token of an originary act of bureaucratic recognition of the 'authentic object'—an 'accurate description' of the bearer recognized and signed by an accredited official, and available for repeated acts of probative ratification." Caplan, "This or That Particular Person," 51. Of course, the question of what constitutes an "accurate description" of the bearer has been an inherent problem of identification systems, one that has led to repeated innovations in ID systems (including the incorporation of photography) and one that facial recognition technology and other biometric technologies promise to resolve definitively.

35. Robertson, "Documentary Regime of Verification," 12.

36. Martin Lloyd, *The Passport: The History of Man's Most Traveled Document* (Thrupp Stroud, UK: Sutton, 2003): 103.

37. Robertson, "Documentary Regime of Verification."

38. Rose, *Powers of Freedom*, 240.

39. Ibid., 246.

40. For David Harvey's explanation of "embedded liberalism," see chapter 1, "Freedom's Just Another Word," in *Brief History of Neoliberalism*, 5–38.

41. Ibid., 10–11.

42. Ibid., 11.

43. Ibid., 12.

44. Ibid., 26.

45. Ibid., 19.

46. Schiller, *Digital Capitalism.* See also Schiller, *How to Think About Information.*

47. Schiller, *Digital Capitalism*, 7.

48. Ibid., 13.

49. Ibid., 14.

50. Andrew Pollack, "Technology: Recognizing the Real You," *New York Times*, September 24, 1981.

51. Ibid.

52. Ibid.

53. Robert Trigaux, "Direct Recognition Joining the Quill; Pen Banks Looking for New Ways to Customer ID—Perhaps Biometric," *American Banker* (October 20, 1982): 29.

54. Ibid.

55. In the 1980s, the banking industry began considering the possibility of appropriating fingerprinting technologies from the realm of criminal identification, and companies developing and marketing these technologies for law enforcement agencies began to reconfigure their design and marketing strategies around employee access control and customer identification applications. In 1986, a company called Identix Inc. introduced the IDX-50, a microprocessor-embedded smart card system equipped with their patented finger imaging technology. Identix priced the IDX-50 system at $7,500 and positioned it as "useful for access-control applications," and for "verification of credit or debit cards." "Biometric System Verifies Identity," *ABA Banking Journal* (April 1986): 84. The Bank of America in Pasadena, California, began using the IDX-50 for controlling employee access to their bank card operations center in 1987. By then Identix also included among its clients Omron Tateisi Electronics, a Japanese manufacturer of ATMs and point-of-sale terminals.

56. Josh Lauer, "From Rumor to Written Record: Credit Reporting and the Invention of Financial Identity in Nineteenth-Century America," *Technology and Culture* 49, no. 2 (2008): 302.

57. Harvey, *Brief History of Neoliberalism*, 33. On the financialization of everyday life, see Randy Martin, *Financialization of Daily Life* (Philadelphia: Temple University Press, 2002).

58. Harvey, *Brief History of Neoliberalism*, 33.

59. Ibid., 159. Emphasis added.

60. Ibid., 159–164.

61. Ibid., 161.

62. Ibid., 159.

63. Robin Stein, "The Ascendancy of the Credit Card Industry," *PBS Frontline* (2004), http://www.pbs.org/wgbh/pages/frontline/shows/credit/more/rise.html.

64. David Evans, *Paying with Plastic: The Digital Revolution in Buying and Borrowing* (Cambridge: MIT Press, 1999).

65. Rule, *Private Lives and Public Surveillance*, 265.

66. David E. Sanger, "Now, 'Smart' Credit Cards," *New York Times*, April 18, 1985.

67. Joseph Turow, *Breaking Up America: Advertisers and the New Media World* (Chicago: University of Chicago Press, 1997): 44.

68. Ibid., 43.

69. Ibid., 44.

70. Ian Hacking, "How Should We Do the History of Statistics?" in *The Foucault Effect: Studies in Governmentality*, ed. Graham Burchell, Colin Gordon, and Peter Miller (Chicago: University of Chicago Press, 1991): 181–196. On the rationalization of consumption, see Kevin Robins and Frank Webster, *Times of the Technoculture: From the Information Society to the Virtual Life* (New York: Routledge, 1999): 98–99. See also James R. Beniger, *The Control Revolution: Technological and Economic Origins of the Information Society* (Cambridge: Harvard University Press, 1986): 344–389.

71. One early adopter of a biometric time management system was the Chicago Housing Authority, which installed a retinal scanning system for this purpose in 1988. The system consisted of forty-seven scanning units that served as time clocks for two thousand housing authority employees, implemented to deal with the problem of "employees who

enlist others to act as stand-ins to sign in and out of housing agency facilities" "Breakthrough for Biometrics," *American Banker* (November 9, 1988): 12.

72. J. W. Toigo, "Biometrics Creep into Business," *Computerworld* (June 11, 1990): 75.

73. For his analysis of the transition from Fordism to the regime of "flexible accumulation," see David Harvey, "Part II: The Political-Economic Transformation of Late Twentieth-Century Capitalism," in *The Condition of Postmodernity* (Malden, MA: Blackwell, 1990): 121–197.

74. Castells, *End of the Millennium*, 72.

75. Robins and Webster, *Times of the Technoculture*, 115.

76. On the history of fingerprinting, see Simon A. Cole, *Suspect Identities: A History of Fingerprinting and Criminal Identification* (Cambridge: Harvard University Press, 2001).

77. U.S. General Accounting Office, *Technology Assessment: Using Biometrics for Border Security* (November 2002), GAO-03-174.

78. According to the developer, Robert "Buzz" Hill, it is not precisely the retina, but "the mat of vessels of the choroid just behind the retina" that provides the most useful source of information for identifying individuals in this method, thus "retinal identification" is somewhat of a misnomer "but nevertheless useful because the term is familiar." Hill, "Retina Identification," in *Biometrics: Personal Identification in a Networked Society*, ed. Anil K. Jain, Ruud Bolle, and Sharath Pankanti (Boston: Kluwer Academic, 1999): 124.

79. Ibid., 130.

80. For a discussion and critical analysis of voiceprint analysis, see Jonathan Sterne, "Enemy Voice," *Social Text 96* 26, no. 3 (2008): 79–100.

81. A 2003 review of the technical literature on facial recognition technology notes that "due to its user-friendly nature, face recognition will remain a powerful tool in spite of the existence of very reliable methods of biometric personal identification such as fingerprint analysis and iris," suggesting that face recognition is not very reliable relative to other biometrics. Zhao et al., "Face Recognition," 453.

82. On signaling theory and the face in mediated contexts, see Judith Donath, "Mediated Faces," in *Proceedings of the 4th International Conference on Cognitive Technology: Instruments of the Mind*, ed. Meurig Beynon, Chrystopher L. Nehaniv, and Kerstin Dautenhahn (London: Springer-Verlag, 2001): 373–390.

83. According to a survey of facial recognition technology in *Biometric Technology Today*, "The reason that facial recognition is so popular is primarily thanks to a vast existing infrastructure of hardware that can use the technology. . . . In particular many existing large-scale public sector ID systems already have acquired facial images, such as for passports, drivers' licences and ID cards." "Facial Recognition—Part 1," *Biometric Technology Today* (November–December 2003): 11.

84. One of the strengths of facial recognition systems, according to *Biometric Technology Today*, is that they "can operate without user knowledge/cooperation." Ibid.

85. Chellappa et al., "Human and Machine Recognition of Faces," 706.

86. On the need to convince stakeholders that a technology is possible, see MacKenzie, *Inventing Accuracy, 384-386.*

87. On using the real world as a laboratory, see ibid., 423.

88. P. Jonathan Phillips, Patrick J. Raus, and Sandor Z. Der, *FERET (Face Recognition Technology) Recognition Algorithm Development and Test Results* (Army Research Laboratory, October 1996), ARL-TR-995, http://www.nist.gov/humanid/feret/doc/army_feret3.

pdf (p. 7). The FERET program focused on three major tasks: establishing the "technology base" necessary for a face recognition system; collecting a large, standardized database of facial images that could be used to develop, test, and evaluate face recognition algorithms; and performing independent, government-monitored testing and evaluation of face recognition algorithms in order to measure their overall progress, determine their level of maturity, and provide independent comparisons. The initial government-supervised tests of facial recognition algorithms were run in August 1994 and March 1995, using images obtained in semi-controlled conditions (with minor lighting and pose variations). The main participants consisted of a select group of researchers chosen based on proposals submitted to the FERET program: the MIT Media Lab, the Rutgers Center for Computer Aids for Industrial Productivity, the USC Computational Vision Lab, and researchers at University of Illinois at Chicago and at Urbana-Champaign, and at the Analytic Science Company. FERET also allowed nonparticipating researchers to access the FERET database and take the FERET tests, but only one additional researcher submitted an algorithm for government testing: Joseph Atick, head of the Laboratory of Computational Neuroscience at Rockefeller University.

89. Ibid., 54

90. Ibid.

91. Eric Convey, "What's Up with That?: Lau to Spin Off ID Division," *Boston Herald*, April 29, 1996.

92. In 1987, Pentland worked on a "people meter" for Arbitron that would identify which family members were actually present in the room watching a television show, rather than depending on their unreliable diaries. Daniel Golden, "The Face of the Future," *Boston Globe*, June 30, 1996.

93. Another way these companies attempted to construct the accuracy and authority of the technology was through the names given to their products and the slogans used to sell them. Here one could discern efforts to define the technology in ways that appealed broadly to investors, institutional users, and the wider public of potential end users and data subjects. At Visionics, for example, Atick and his colleagues named their facial recognition product "FaceIt®" and spun it as an "Enabling Technology with Mass Appeal." The name "FaceIt" (always accompanied in print by the trademark symbol "®"), for example, circulated in news reporting and other public discourse as an identifying term for the Visionics facial recognition product; though not quite a household word, it became one of the more prominent product names associated with the technology. A loaded signifier and curious identifier for the technology, the term "FaceIt" begs some interpretative unpacking. Reading "It" as an acronym, "FaceIt" might be interpreted as a variation of "Face IT," short for "face information technology" or "face identification technology." Read as an immediate address, "FaceIt" suggests that subjects should literally face the technology, or face the camera, so that their bodies can be identified. On a connotative level, the phrase "face it" signifies a demand to "face facts" or accept the truth, which begs the question as to what facts or truth are being asserted. Were people being asked to accept the technology's authority to accurately bind identity to the body? Should they face the inevitability of the technology's widespread diffusion? Perhaps "it" was meant as a hopeful reference to a utopian technological future that should be embraced rather than resisted. Of course, it is not clear exactly what facts are being asserted, since the pronoun "it" is without a referent. In this sense, "FaceIt" seemed to invoke generalized

insecurities about the present and the future, precisely the justifications asserted for new and improved surveillance and identification technologies. "FaceIt" is just the sort of meaningless phrase that Orwell might have taken to task in "Politics and the English Language," or incorporated into the fictional newspeak vocabulary as shorthand for a more complex statement no longer accessible. The obvious Orwellian underpinnings of "FaceIt" make it a curious and almost consciously ironic choice to identify a facial recognition product.

The slogan "Enabling Technology with Mass Appeal" is similarly ironic and without actors or subjects, generating vague connotations of empowerment and popular enthusiasm. Exactly *what* would "FaceIt" enable and what "mass" would it appeal to? The adjective "enabling" signifies making something possible; that which enables lends agency to something or someone else. But there was no precise subject or actor being enabled in this phrase; so while the vague suggestion was one of individual empowerment, it could just as easily mean enabling the state to track the movement of people, or enabling organizations to monitor their employees' workday. The technology ostensibly enables other technologies to function in more effective ways; for example, the company claims that FaceIt makes "passive," closed-circuit television systems more "active" by automatically scanning video feeds, extracting faces, converting them into digital templates, and searching them against a watchlist database. In this sense, it promises to "enable" more effective use of CCTV systems and improved practices for state and private security agencies that employ them. The other odd choice of terms in the sales slogan—"mass appeal"—invokes an image of hordes of people clamoring to use the technology. Such an image detracts from the idea of *individual* identification, since a "mass" is the opposite of specific individuals. Ironically, "mass appeal" also signifies manipulation and control, leaving an appropriate if unintentional impression.

94. "Miros Intros 'Computer Face Recognition' Technology," *Newsbytes*, June 9, 1994.

95. Christine L. Lisetti and Diane J. Schiano, "Automatic Facial Expression Interpretation," *Pragmatics and Cognition* 8, no. 1 (2000): 203.

96. Alexandra Stikeman, "Biometrics: Joseph Atick" *Technology Review* (February 1, 2003), http://www.technologyreview.com/InfoTech/12255.

97. State agencies have been centrally involved in biometric system development on many different levels. The Defense Department has of course provided significant subsidies for research and development. The National Security Agency has also been involved, for example, sponsoring a collaborative industry-government association called the Biometrics Consortium, which has worked heavily in the area of technology standards. The National Institute of Standards and Technology has had a hand in developing biometrics standards as well, and DARPA and other federal government agencies have sponsored testing of commercial biometric systems. These tests are very expensive and invaluable to industries trying to determine which biometric systems to adopt for their needs. And of course the state provides the regulatory framework for biometric system deployment, whether policies mandating the adoption of more secure forms of identification, or lax privacy policies that allow corporations to own and control data about consumers. The industry "self-regulatory" model of privacy has also given employers carte blanche to monitor everything workers do in the workplace.

98. Quoted in Rex Crum, "The Latest in ID Technology: It's Written All Over Your Face," *Boston Business Journal* 18, no. 1 (1998): 13.

99. "New Mexico Puts Digital Images on Drivers' Licenses," *Newsbytes*, April 21, 1997; Polaroid Corporation, "Polaroid and Visionics Team Up to Provide Facial Recognition Solutions from Department of Motor Vehicles," *PR Newswire*, April 15, 1998.

100. John Torpey, *The Invention of the Passport: Surveillance, Citizenship, and the State* (Cambridge: Cambridge University Press, 2000).

101. According to a U.S. General Accounting Office report, by 2001 the Department of Defense had supported facial recognition technology research to the tune of $24.7 million, and the National Institute of Justice had spent $21.3 million. U.S. General Accounting Office, *Federal Funding for Selected Surveillance Technologies* (March 14, 2002), GAO-02-438R, http://www.gao.gov/new.items/d02438r.pdf.

102. Nils Christie, *Crime Control as Industry: Towards Gulags, Western Style*, 3rd ed. (New York: Routledge, 2000). There are other connections between law enforcement interest in biometrics and the war on drugs. The first piece of federal legislation to include biometrics provisions was the Anti–Drug Abuse Act of 1988, which required the establishment of minimum standards for the use of biometrics to identify commercial motor vehicle operators. Also, in addition to sponsoring testing and evaluation of commercial facial recognition systems, the Defense Counterdrug Technology Development Office purchased the Visionics FaceIt facial recognition surveillance system in December 2001.

103. Ibid.

104. William Welsh, "Florida County Gives Viisage $2.7 Million Award," *Newsbytes*, August 2, 2001.

105. Visionics Corporation, "State of Minnesota Adopts Visionics' FaceIt Technology for Integrated Criminal Justice Mug Shot Database System," August 14, 2001, http://lone.client.shareholder.com/releasedetail.cfm?releaseid=208878.

106. Aihwa Ong defines neoliberalism as "a new relationship between government and knowledge through which governing activities are recast as nonpolitical and nonideological problems that need technical solutions." Ong, *Neoliberalism as Exception* (Durham: Duke University Press, 2006): 3.

107. Harvey, *Brief History of Neoliberalism*, 77.

108. Ibid., 165.

109. Loïc Wacquant, "The Penalisation of Poverty and the Rise of Neo-Liberalism," *European Journal on Criminal Policy and Research* 9, no. 4 (2001): 401.

110. Ibid.

111. Jonathan Simon, *Governing Through Crime: How the War on Crime Transformed American Democracy and Created a Culture of Fear* (Oxford: Oxford University Press, 2007): 152–153.

112. As is widely recognized, African American men have shouldered the burden of these trends, being heavily overrepresented among the exploding prison population.

113. Dana Milbank, "Measuring and Cataloging Body Parts May Help to Weed Out Welfare Cheats," *Wall Street Journal*, December 4, 1995.

114. Ibid.

115. For example, the Massachusetts Department of Transitional Assistance adopted facial recognition technology from Viisage Technology in 1996. Pam Derringer, "Welfare Cards Picture Perfect," *Mass High Tech*, April 29, 1996.

116. Polaroid Corporation, "Polaroid and Visionics Team Up to Provide Facial Recognition Solutions from Department of Motor Vehicles," *PR Newswire*, April 15, 1998.

117. "Polaroid and Visionics Tout Face Recognition," *Newsbytes*, April 16, 1998.

118. Shoshana Magnet, "Bio-benefits: Technologies of Criminalization, Biometrics, and the Welfare System," in *Surveillance: Power, Problems, and Politics*, ed. Sean Hier and Joshua Greenberg (Vancouver: University of British Columbia Press, 2009): 169–183.

119. Peter Andreas, "Introduction: The Wall After the Wall," in *The Wall Around the West: State Borders and Immigration Controls in North America and Europe*, ed. Peter Andreas and Timothy Snyder (Lanham, MD: Rowman and Littlefield, 2000): 4.

120. The integration of biometrics into state identification systems for border and immigration control began in the early 1990s with the development of IDENT, the biometric fingerprint system of the INS. IDENT was envisioned to deal with the "revolving door" phenomenon of undocumented entry into the United States by workers from south of the U.S.-Mexico border. While IDENT was primarily conceived as a fingerprint system, the State Department has also begun to build a database of digital photographs of those apprehended.

121. Giorgio Agamben refers to individuals who fall outside the realm of legal rights as *homo sacer*. See Agamben, *Homo Sacer: Sovereign Power and Bare Life*, trans. Daniel Heller-Roazen (Palo Alto: Stanford University Press, 1998).

122. James Boyle discusses the prevalence of the "author paradigm" in the way we think about and regulate information: "The author-vision conjures up a new political economy of wealth supported, and reflexively constituted, by a particular ideology of entitlement. At the bottom of the pyramid would be those whose lives, bodies, and natural environments provide the raw materials, or those who themselves are the ultimate 'audience,' for the products of an information economy. At the top of the pyramid of entitlement claims—at least in theory—would be those who gather and shape information and information products. More important, perhaps, is the fact that the bureaucratic and corporate actors who *employ* the manipulators of information can justify their own, derivative intellectual property rights through the rhetoric of individualism and original genius. In fact, a striking feature of the language of romantic authorship is the way it is used to support sweeping intellectual property rights for large corporate entities." Boyle, *Shamans, Software, and Spleens: Law and the Construction of the Information Society* (Cambridge: Harvard University Press, 1996): xii–xiii.

123. Haggerty and Ericson, "Surveillant Assemblage," 606.

124. Foucault, *Discipline and Punish*, 144.

125. In order to address this obstacle to full-scale biometric adoption, the biometrics industry, along with other IT developers, began to cooperate and organize around developing standards, working on Application Programming Interfaces (APIs) that could be incorporated into operating systems and application software. In 1998, Compaq, IBM, Identicator Technology, Microsoft, Miros, and Novell formed the BioAPI Consortium to pursue biometrics APIs. The development of standards is never a neutral process, and can perform a sort of self-fulfilling prophecy in the trajectory of technological development, as Geoffrey Bowker and Leigh Star have noted. Once established, "standards have significant inertia and can be very difficult and expensive to change." Bowker and Star, *Sorting Things Out: Classification and Its Consequences* (Cambridge: MIT Press, 2000): 14.

126. The development of CBIR techniques would also require partnerships with other IT companies that had a mutual interest in designing more sophisticated forms of image retrieval. For example, in 1997 Viisage partnered with the leading database software

firm, Oracle Corporation, and Virage Inc., a developer of image and video processing systems, to integrate the Viisage technology with Oracle's new Oracle8 database system, attempting to make "the ability to identify and verify faces a natural part of the database." Viisage Technology Inc., "Viisage Technology Inc. Adds Facial Recognition to Oracle8," *PR Newswire*, June 24, 1997. As we will see in chapter 4, developers have also worked to make the ability to identify and verify faces a natural part of online photo management, enlisting Internet users to experiment with prototype facial recognition search engines in order to analyze and improve CBIR techniques.

127. Another border inspection program, called the Secure Electronic Network for Travelers Rapid Inspection (SENTRI), experimented with facial recognition technology from Visionics. SENTRI was implemented at the Otay Mesa, California, border-crossing station in October 1995, and was designed as an automated commuter lane that would expedite the crossing of certain low risk, pre-enrolled border crossers, using Automatic Vehicle Identification (AVI). The SENTRI project experimented with Visionics facial recognition technology in 1997. According to biometrics eexpert James Wayman of San Jose State University, however, the facial recognition system did not work properly because of the varying heights of cars and because the window frames obscured people's faces. Lev Grossman, "Welcome to the Snooper Bowl," *Time* (February 12, 2001): 72.

128. Major deficiencies in the BCC system came to light in congressional hearings held after September 11, 2001. See "Using Information Technology to Secure America's Borders: INS Problems with Planning and Implementation," Hearing of the Immigration and Claims Subcommittee of the House Judiciary Committee, 107th Cong. (October 11, 2001). See also "Biometric Identifiers and the Modern Face of Terror: New Technologies in the Global War on Terrorism," Hearing Before the Technology, Terrorism, and Government Information Subcommittee of the Judiciary Committee, U.S. Senate, 107th Cong. (November 14, 2001).
Both IDENT and the BCC were fraught with logistical, technical, funding, and other problems. A review of IDENT following the highly publicized Resendez-Ramirez murder case in 1999 revealed, among other things, that INS officers lacked training on the system, that fingerprint scanners were broken or otherwise not operational, and that a fully functional IDENT system "would require an additional $400 to $700 million in staffing costs to handle the increased workload resulting from the identification of apprehended aliens . . . *plus* an additional $200 to $700 million in detention costs for the increased number of aliens who would have to be detained" as criminals according to federal law. U.S. Department of Justice, *Status of IDENT/IAFIS Integration* (December 7, 2001), I-2002-003, http://www.justice.gov/oig/reports/plus/e0203/finger.htm. As a result of these and other problems, Congress directed the INS to suspend further deployment of IDENT until there was a plan in place to integrate IDENT with the FBI's criminal fingerprint system (Integrated Automated Fingerprint Identification System, or IAFIS). U.S. Department of Justice, *The Rafael Resendez-Ramirez Case: A Review of the INS's Actions and the Operation of Its IDENT Automated Fingerprint Identification System* (March 20, 2000), http://www.justice.gov/oig/special/0003/index.htm.

129. Brian Winston uses the term "supervening social necessities" to refer to the accelerators transforming prototype technologies into widespread use. Conversely, the "law of suppression of radical potential" refers to a concentration of social constraints serving as "brakes" and limiting the disruptive effects of technologies on existing social formations.

Winston, *Media Technology and Society: A History: From the Telegraph to the Internet* (New York: Routledge, 1998): 11–13.

130. Hill, "Retina Identification," 130.

131. Harvey, *Condition of Postmodernity*, 123.

132. Stikeman, "Technology Review Top Ten: Biometrics."

## NOTES TO CHAPTER 2

1. Debra Seagal, "Tales from the Cutting-Room Floor: The Reality of Reality-Based Television," *Harper's Magazine* (November 1993): 50.

2. Ibid., 55.

3. Clive Norris, Jade Moran, and Gary Armstrong, "Algorithmic Surveillance: The Future of Automated Visual Surveillance," in *Surveillance, Closed Circuit Television, and Social Control*, ed. Norris, Moran, and Armstrong (Aldershot, UK: Ashgate, 1998): 256.

4. On "algorithmic surveillance" see ibid. See also Introna and Wood, "Picturing Algorithmic Surveillance." As Graham and Wood have argued, there are intimate connections between the digitization of police surveillance techniques and the changing political economy of cities. Graham and Wood, "Digitizing Surveillance." Still, low-wage labor may turn out to be less expensive than algorithmic or digitized forms, and in fact the World Bank has proposed employing workers in Africa to monitor CCTV systems in U.S. shopping malls. Michael McCahill, "Beyond Foucault: Towards a Contemporary Theory of Surveillance," in *Surveillance, Closed Circuit Television, and Social Control*, ed. Clive Norris, Jade Moran, and Gary Armstrong (Aldershot, UK: Ashgate, 1998): 45. On surveillance labor, see Richard Maxwell, "Surveillance: Work, Myth, and Policy," *Social Text 83* 23, no. 2 (2005): 1–19.

5. James Boyle uses the term "digital enclosure" to refer to the process of expanding intellectual property rights in information—that is, a new enclosure movement akin to the land enclosure movement in England, whereby common lands used for pastures and farming were fenced off for private use, creating a landless working class and dramatically altering social and economic life. Boyle, "The Second Enclosure Movement and the Construction of the Public Domain," *Law and Contemporary Problems* 66 (Winter–Spring 2003): 33–74. Mark Andrejevic uses the term "digital enclosure" to refer to "the creation of an interactive realm wherein every action and transaction generates information about itself," a process "whereby places and activities become encompassed by the monitoring embrace of an interactive (virtual) space." Andrejevic, *iSpy*, 2. See also Andrejevic, "The Work of Being Watched: Interactive Media and the Exploitation of Self-Disclosure," *Critical Studies in Media Communication* 19, no. 2 (2002): 238.

6. Anna McCarthy, "Closed Circuit Television," Museum of Broadcast Communications (2010), http://www.museum.tv/archives/etv/C/htmlC/closedcircui/closedcircui.htm.

7. Ibid. In the earlier days of television, CCTV was envisioned for a wide range of uses that today are the domain of distributed, packet-switched computer networks. In *Mass Communications and American Empire*, Herbert Schiller cites a 1965 White House Conference on International Communication as envisioning the "use of closed circuit television by major companies doing a worldwide business to link together their offices and affiliates in order to kick off a new sales campaign or to demonstrate a new product or to analyze a new marketing situation in the far corners of the world." Schiller, *Mass Communications and American Empire*, 2nd ed. (Boulder: Westview Press, 1992): 148.

8. According to Chris Williams, the fact that police in the United Kingdom were using CCTV as far back as the 1960s raises questions about prevailing theoretical associations between police use of these systems and the rise of "postindustrial society." Williams, "Police Surveillance and the Emergence of CCTV in the 1960s," *Crime Prevention and Community Safety* 5, no. 3 (2003): 27.

9. Clive Norris, Michael McCahill, and David Wood have highlighted some of the important differences between countries with regard to the extent and pace of CCTV deployment, and argue that "the complex interplay of local and regional culture and politics around surveillance technologies in diverse cultures needs further investigation." Norris, McCahill, and Wood, "The Growth of CCTV: A Global Perspective on the International Diffusion of Video Surveillance in Publicly Accessible Space," *Surveillance and Society* 2, no. 2 (2004): 127.

10. Ibid., 114.

11. Christopher Wilson, *Cop Knowledge: Police Power and Cultural Narrative in Twentieth-Century America* (Chicago: University of Chicago Press, 2000): 5.

12. Stuart Hall, Chas Critcher, Tony Jefferson, John Clarke, and Brian Roberts, *Policing the Crisis: Mugging, the State, and Law and Order* (London: Palgrave, 1978).

13. Terry McCarthy, "The Gang Buster," *Time* (January 19, 2004): 56.

14. Wilson, *Cop Knowledge*, 10.

15. David Garland, *The Culture of Control: Crime and Social Order in Contemporary Society* (Chicago: University of Chicago Press, 2001); Garland, "The Limits of the Sovereign State: Strategies of Crime Control in Contemporary Society," *British Journal of Criminology* 36, no. 4 (1996): 445–471.

16. Garland, "Limits of the Sovereign State," 455.

17. Garland, *Culture of Control*, 122.

18. Wilson, *Cop Knowledge*, 170.

19. Aaron Doyle has argued that the practice of broadcasting surveillance footage of crimes itself has encouraged "a more emotionally charged, punitive, vengeance-oriented approach to crime and control." Doyle, "An Alternative Current in Surveillance and Control: Broadcasting Surveillance Footage of Crimes," in *The New Politics of Surveillance and Visibility*, ed. Kevin D. Haggerty and Richard V. Ericson (Toronto: University of Toronto Press, 2006): 201.

20. Stephen Graham, "Towards of Fifth Utility? On the Extension and Normalisation of Public CCTV," in *Surveillance, Closed Circuit Television, and Social Control*, ed. Clive Norris, Jade Moran, and Gary Armstrong (Aldershot, UK: Ashgate, 1998): 89.

21. See, for example, Michael McCahill, *The Surveillance Web: The Rise of Visual Surveillance in an English City* (Portland: Willan, 2002); Clive Norris and Gary Armstrong, *The Maximum Surveillance Society* (Oxford: Berg, 1999).

22. McCahill, *Surveillance Web*, 99.

23. Lynsey Dubbeld, "The Role of Technology in Shaping CCTV Surveillance Practices," *Information, Communication, and Society* 8, no. 1 (2005): 88.

24. Richard V. Ericson and Kevin D. Haggerty, *Policing the Risk Society* (Toronto: University of Toronto Press, 1997): 389–390. In his study of the relationship between information technology and the organizational structure of policing, Peter Manning similarly emphasizes the symbolic role of new technologies: "Technology is a ploy in games of power and control within organizations, a way of amplifying uncertainty, and

a source of symbolic capital for management, independent of its usefulness in achieving stated public goals or objectives." Manning, "Information Technology and the Police," in *Crime and Justice* 15, *Modern Policing*, ed. Michael H. Tonry and Norval Morris (Chicago: University of Chicago Press, 1992): 391.

25. On the major limitations and failures of CCTV systems to fulfill their original objectives of crime prevention and public safety, see Nic Groombridge, "Stars of CCTV? How the Home Office Wasted Millions—A Radical 'Treasury/Audit Commission' View," *Surveillance and Society* 5, no. 1 (2008): 73-80; Leon Hempel and Eric Töpfer, "The Surveillance Consensus: Reviewing the Politics of CCTV in Three European Countries," *European Journal of Criminology* 6, no. 2 (2009): 157–177; David Murakami Wood, "A New 'Baroque Arsenal'? Surveillance in a Global Recession," *Surveillance and Society* 6, no. 1 (2009): 1-2; and William Webster, "CCTV Policy in the UK: Reconsidering the Evidence Base," *Surveillance and Society* 6, no. 1 (2009): 10–22.

26. Sekula, "The Body and the Archive."

27. "FBI Setting Standards for Computer Picture File of Criminals," *New York Times*, November 5, 1995.

28. Industry consolidation among companies marketing commercial facial recognition systems in the early 2000s confirmed and necessitated that they would continue to seek big-budget contracts with institutional users seeking technologies to support large-scale identification and access-control systems. In February 2001, Visionics Corporation merged with Digital Biometrics, a fingerprint technology vendor selling primarily to the law enforcement market, although also positioning their technologies as valuable to commercial employers and government agencies for background-checking employment and permit applicants. By merging with Digital Biometrics, Visionics diversified its biometric product offerings and moved toward developing and selling systems that would incorporate more than one type of biometric technology. Then in the summer of 2002, Visionics merged with Identix Inc., another leading optical fingerprint company. Founded in 1982, Identix purchased Visionics for $226.9 million, creating a combined company worth $600 million. See Identix Inc., "Identix and Visionics Agree to Merge, Creating World's Leading Multi-biometric Security Technology Company" (February 22, 2002), http://ir.l1id.com/releasedetail.cfm?ReleaseID=208845. Joseph Atick became CEO of the combined company, assuring stockholders and potential investors that he would "be relentless in capitalizing on the tremendous opportunity before us" to offer "total security solutions" and leverage "the technologies and partnerships of each company into key large-scale markets" (ibid). Also in the summer of 2002, Viisage acquired Miros Inc., and soon thereafter Viisage merged with the German vendor ZN Vision Technologies, creating the world's largest facial recognition company. At the time of the merger, Viisage was 31 percent owned by U.S. defense contractor Lau Technologies, and its customers included Motor Vehicle departments in about twenty U.S. states and the U.S. Immigration and Naturalization Service. ZN Vision Technologies had 170 institutional clients, mostly in Europe. Viisage also acquired Digital Technologies Corporation, a privately held firm that serviced the production of U.S. passports using a patented digital printing system. See Viisage Technology, "Viisage Technology to Acquire ZN Vision Technologies AG," *Business Wire*, March 31, 2003.

Then in 2006, Viisage and Identix, the two leading competitors, merged under the auspices of a new company called L-1 Identity Solutions. According to the press release, the merger would establish "the industry's most comprehensive multi-modal biometric platform," thereby "unlocking the potential of strong synergies." L-1 Identity Solutions, "Viisage and Identix to Merge to Create Biometric Identity Solution Leader" (January 12, 2006), http://lone.client.shareholder.com/releasedetail.cfm?ReleaseID=209035. Through mergers, these companies pooled their patents, combined development and marketing efforts, and no longer competed for the same customers. In terms of the technology itself, merging the companies allowed for a fusion of techniques and the further development of multimodal applications, combining multiple biometrics into one system to suit each user's specific needs. Although there were other players in the facial recognition technology industry by 2006, the merger of Viisage and Identix produced a company with a large chunk of the market, providing facial recognition and other biometric systems to a critical mass of customers (often through system integrators and original equipment manufacturers). The "strong synergies" created by consolidation among facial recognition companies has meant more centralized control of the design of large-scale identification systems. Scaled-up companies like L-1 Identity Solutions have a hand in implementing a wide range of systems, potentially offering a greater degree of technical compatibility among different systems. The success of the "strong synergies" that are "unlocked" by corporate consolidation of biometric companies is a measure of the broadening scale and reach of the emerging "surveillant assemblage."

29. Duane M. Blackburn, Mike Bone, and P. Jonathan Phillips, *Facial Recognition Vendor Test 2000: Executive Overview* (December 2000), http://www.frvt.org/DLs/FRVT2000_Executive_Overview.pdf.

30. Ibid.

31. Visionics Corporation, "Visionics Demonstrates World's First Face Recognition Engine Capable of Identifying and Tracking Multiple Faces," *PR Newswire*, October 7, 1997.

32. Visionics Corporation, "Find Criminals, Missing Children, Shoplifters, Terrorists in a Crowd Using Face Recognition Software Linked to a Database," *PR Newswire*, March 10, 1998.

33. Visionics Corporation, "Government Awards London Borough with Major Grant to Expand CCTV Surveillance Program," *Business Wire*, February 7, 2000.

34. Visionics Corporation, "Visionics Corporation Announces FaceIt ARGUS—First Commercially Available Scalable Solution for Real Time Facial Surveillance and Identification," *Business Wire*, October 1, 2001. Emphasis added.

35. I would like to thank Jonathan Sterne for suggesting this line of inquiry concerning facial recognition technology as a so-called laborsaving device that embodies the logic of interpassivity. The way that Žižek uses concept of interpassivity is reminiscent of Adorno's argument that popular music listens for us, in that it is formulaic and requires no active engagement on the part of the listener. Slavoj Žižek, *The Plague of Fantasies* (New York: Verso): 111–113.

36. Gijs Van Oenen, "A Machine That Would Go of Itself: Interpassivity and Its Impact on Political Life," *Theory and Event* 9, no. 2 (2006), http://muse.jhu.edu/journals/theory_and_event/v009/9.2vanoenen.html.

37. "Terrorism Prevention: Focusing on Biometric Identifiers," Hearing Before the Technology, Terrorism, and Government Information Subcommittee of the Judiciary Committee, U.S. Senate. 107th Cong. (November 14, 2001), prepared testimony of Joseph Atick.

38. In Richard Thomas, "As UK Crime Outstrips the US, a Hidden Eye Is Watching," *The Observer*, October 11, 1998.

39. Robert Lack, "Development of Facial Recognition Technologies in CCTV Systems," *The Source Public Management Journal*, October 25, 1999.

40. Robert Lack, "Use of CCTV and Development of Facial Recognition Technologies in Public Places," paper presented at the International Conference of Data Commissioners, September 24, 2001. Emphasis added.

41. Thomas, "As UK Crime Outstrips the US, a Hidden Eye Is Watching."

42. Pat Oldcorn, quoted in ibid.

43. Watkins ran his own company, Advanced Biometric Imaging, which later became ATC International. His company also installed a police Smart CCTV system in Virginia Beach in the summer of 2002.

44. David Watkins, personal communication, August 23, 2003.

45. Visionics Corporation, "Tampa Police Department Installs Visionics' FaceIt Technology in Anti-crime CCTV Initiative," *Business Wire*, June 29, 2001.

46. Gary R. Mormino and George E. Pozzetta, *The Immigrant World of Ybor City: Italians and Their Latin Neighbors in Tampa, 1885–1985* (Urbana: University of Illinois Press, 1987).

47. Ibid.

48. Ibid., 306.

49. Ibid.

50. Eric Snider, "This Party's Dying," *Creative Loafing*, March 12, 2003.

51. Bernice Stengle, "Ybor City: 'The Ball's Going to Roll Now,'" *St. Petersburg Times*, February 28, 1988.

52. Alex Sanchez, "Ybor City's Past Can Enhance Tampa's Future," *St. Petersburg Times*, November 24, 1991.

53. "About Ybor City Development Corporation Inc," http://www.tampagov.net/dept_ybor_city_development_corporation/about_us/index.asp.

54. Ybor City Development Corporation, "Plans and Studies," http://www.tampagov.net/dept_ybor_city_development_corporation/information_resources/development_information/plans_and_studies.asp.

55. Timothy Simpson, "Communication, Conflict, and Community in an Urban Industrial Ruin," *Communication Research* 22, no. 6 (1995): 702–703.

56. Snider, "This Party's Dying."

57. Ibid.

58. Ibid.

59. David Harvey, "Flexible Accumulation Through Urbanization: Reflections on 'Post-modernism' in the American City," in *Post-Fordism: A Reader*, ed. Ash Amin (Cambridge, MA: Blackwell, 1994): 361–386.

60. Michael Sorkin, "Introduction: Variations on a Theme Park," in *Variations on a Theme Park: The New American City and the End of Public Space*, ed. Sorkin (New York: Hill and Wang, 1992): xi–xv.

61. As Susan Christopherson has argued, the close relationship between contemporary spaces of consumption and the celebration of cultural diversity and multiculturalism is one of the more interesting twists in the contemporary redesign of urban spaces. In gentrifying urban neighborhoods, the conflation of a "multicultural experience" with

consumption reflects "a struggle between genuine ethnic culture and that which is manufactured for sale. This commodified version of diversity is not about traditions and needs but about surfaces—colours, styles, tastes, all packaged in easily consumable forms." Christopherson, "The Fortress City: Privatized Spaces, Consumer Citizenship," in *Post-Fordism: A Reader*, ed. Ash Amin (Cambridge, MA: Blackwell, 1994): 414. What makes this commodified, stylized "Latinization" of U.S. cities more insidious is that it has expedited the displacement of Latino populations, as Arlene Dávila has shown in *Barrio Dreams: Puerto Ricans, Latinos, and the Neoliberal City* (Berkeley: University of California Press, 2004).

62. Mike Davis, *City of Quartz* (New York: Vintage, 1992): 223, 224.

63. Ivan J. Hathaway, "Surveillance Cameras to Focus on Ybor," *Tampa Tribune*, July 22, 1997.

64. Richard Danielson, "Police Cameras May Soon Scan Ybor," *St. Petersburg Times*, May 24, 1996.

65. Hathaway, "Surveillance Cameras to Focus on Ybor."

66. Ibid.

67. Editorial, "An Alarming Crime in Ybor City," *Tampa Tribune*, June 12, 1996.

68. Richard Hunter, *World Without Secrets: Business, Crime, and Privacy in the Age of Ubiquitous Computing* (New York: Gartner, 2002).

69. Quoted in ibid., 20.

70. Daniel J. Czitrom, *Media and the American Mind: From Morse to McLuhan* (Chapel Hill: University of North Carolina Press, 1982): 114.

71. Rose, *Powers of Freedom*, 72.

72. Ibid., 73.

73. Ibid.

74. Michel Foucault, *Security, Territory, Population: Lectures at the College de France, 1977–78*, ed. Michel Senellart, trans. Graham Burchell (London: Palgrave, 2007): 18.

75. Michel Foucault, *Discipline and Punish*, 201.

76. Here I am using the term "digital enclosure" according to Mark Andrejevic's definition. Andrejevic, *iSpy*, 2.

77. There are applications of facial recognition that are not database driven, and instead involve simply matching a face in an existing image to an image taken of live person. For example, local feature analysis could be used to compare a photo ID to an image taken of the person in *possession* of the ID, confirming that the live person is the same individual whose picture is embedded in the document. This involves a simple one-to-one match without recourse to a database of images. But even if this type of application were most common, a quantity of facial images was nevertheless needed to develop the technique, in order to determine what statistical patterns were characteristic of faces. This is more clearly the case with the eigenface technique, but it is true of all algorithmic techniques of facial recognition. As feminist and other critiques of medical research have demonstrated, the selection of human subjects in research practice is not irrelevant to the outcomes of research.

78. See the press release announcing the system's installation, Visionics Corporation, "Tampa Police Department Installs Visionics' FaceIt Technology in Anti-crime CCTV Initiative," *Business Wire*, June 29, 2001.

79. Ibid.

80. Bruno Latour, *Aramis, or the Love of Technology*, trans. Catherine Porter (Cambridge: Harvard University Press, 1996): 172.

81. David McGuire, "Rep. Armey Blasts Tampa Over Face-Recognition System," *Newsbytes*, July 2, 2001.

82. Quoted in Dana Canedy, "Tampa Scans the Faces in Its Crowds for Criminals," *New York Times*, July 4, 2001.

83. "Surveillance Cameras Incite Protest," *New York Times*, July 16, 2001; Martin Kasindorf, "'Big Brother' Cameras on Watch for Criminals." *USA Today*, August 2, 2001.

84. Mike Meek, "You Can't Hide Those Lying Eyes in Tampa," *U.S. News and World Report* (August 6, 2001), 20.

85. Quoted in Angela Moore, "Snoopy Software Has Council Foe," *St. Petersburg Times*, July 4, 2001.

86. Quoted in Laura Kinsler, "Greco Says Ybor City Cameras Will Stay," *Tampa Tribune*, July 25, 2001.

87. Ibid.

88. Quoted in Meek, "You Can't Hide Those Lying Eyes in Tampa," 20.

89. Quoted in Canedy, "Tampa Scans the Faces in Its Crowds for Criminals."

90. Bob Buckhorn, personal communication, August 22, 2002. Emphasis added.

91. Quoted in Meek, "You Can't Hide Those Lying Eyes in Tampa," 20.

92. Patricia A. Benton, letter to the editor, *Tampa Tribune*, July 14, 2001.

93. Corbett, quoted in Thomas, "As UK Crime Outstrips US, a Hidden Eye Is Watching."

94. Garland, *Culture of Control*.

95. Ibid., 10.

96. Doreen Massey, *Space, Place, and Gender* (Minneapolis: University of Minnesota Press, 1994): 168.

97. Wilson, *Cop Knowledge*, 217.

98. Garland, *Culture of Control*, 12.

99. Rose, *Powers of Freedom*, 70.

100. Ibid.

101. Ibid.

102. The ACLU report indicated that, after three requests, the police furnished "copies of police logs filled in by system operators between July 12 and August 11, 2001; a Memorandum of Understanding between the software manufacturer, Visionics, and the police department; the department's Standard Operating Procedure governing the use of the system; and a training video." Jay Stanley and Barry Steinhardt, *Drawing a Blank: The Failure of Facial Recognition Technology in Tampa, Florida* (American Civil Liberties Union, January 3, 2002), https://www.aclu.org/FilesPDFs/drawing_blank.pdf.

103. Brady Dennis, "Ybor Cameras Won't Seek What They Never Found," *St. Petersburg Times*, August 20, 2003.

104. Quoted in Thomas W. Krause, "City Unplugs Camera Software," *Tampa Tribune*, August 20, 2003.

105. Quoted in Mitch Stacy, "Tampa Police Eliminate Facial-Recognition System," *Atlanta Journal-Constitution*, August 21, 2003.

106. Quoted in Dennis, "Ybor Cameras Won't Seek What They Never Found."

107. Quoted in Stacy, "Tampa Police Eliminate Facial-Recognition System."

108. Quoted in Krause, "City Unplugs Camera Software."

109. There were no significant drops in the crime rate in Ybor City during the experiment. Tampa crime statistics by area are available at the Tampa Police website, http://www.tampagov.net/dept_Police/information_resources/Crime_Statistics/annual_crime_report_booklets.asp. Ybor City is located in grid numbers 127 and 128. Grid 128 remained the highest crime grid in Tampa in 2001 and 2002, the two years during which the Smart CCTV experiment took place.

## NOTES TO CHAPTER 3

1. Hobsbawm, *Age of Extremes.*

2. Seth Grimes, "Shared Risk, Shared Rewards," *Intelligent Enterprise* (September 1, 2003): 28. Robert O'Harrow Jr. makes the argument that what has emerged in the post–9/11 context is a "surveillance-industrial complex" much like the military-industrial complex, with profit motives fueling widespread deployment of new surveillance technologies and systems. O'Harrow, *No Place to Hide: Behind the Scenes of Our Emerging Surveillance Society* (New York: Free Press, 2005). A large volume of literature addresses the proliferation of surveillance practices after September 11, 2001. See, for example, Ball and Webster, eds., *Intensification of Surveillance*; David Lyon, *Surveillance After September 11* (Cambridge: Polity, 2003); Kevin D. Haggerty and Amber Gazso, "Seeing Beyond the Ruins: Surveillance as a Response to Terrorist Threats," *Canadian Journal of Sociology* 30, no. 2 (2005): 169–187.

3. Paul Kaihla, "The Hope of Technology: Stopping the Next One," *Business 2.0* (November 2001): 98.

4. Bruno Latour, *Science in Action: How to Follow Scientists and Engineers Through Society* (Cambridge: Harvard University Press, 1987): 1–17.

5. Michael R. Curry makes a related argument about the use of computer-based geodemographic systems for profiling travelers: notwithstanding the contention that they are more objective than earlier airline profiling systems, computer-based models persist in relying on stereotypical narratives to make determinations about "treacherous" versus "trusted" travelers. Curry, "The Profiler's Question and the Treacherous Traveler: Narratives of Belonging in Commercial Aviation," *Surveillance and Society* 1, no. 4 (2004): 476. For additional social critiques of profiling technologies, see Greg Elmer, *Profiling Machines: Mapping the Personal Information Economy* (Cambridge: MIT Press, 2004); and Gandy, *Panoptic Sort*. For additional analyses of the securitization of identity in the post-9/11 context, see Benjamin J. Muller, "(Dis)qualified Bodies: Securitization, Citizenship, and 'Identity Management,'" *Citizenship Studies* 8, no. 3 (2004): 279–294; Kim Rygiel, "Protecting and Proving Identity: The Biopolitics of Waging War Through Citizenship in the Post-9/11 Era," in *(En)gendering the War on Terror: War Stories and Camouflaged Politics*, ed. Krista Hunt and Kim Rygiel (Aldershot, UK: Ashgate, 2006): 146–167; Mark B. Salter, "When the Exception Becomes the Rule: Borders, Sovereignty, and Citizenship," *Citizenship Studies* 12, no. 4 (2008): 365–380.

6. John Poindexter, "Finding the Face of Terror in Data," *New York Times*, September 10, 2003.

7. On the politics of "global information dominance," see Armand Mattelart, *The Information Society* (Thousand Oaks, CA: Sage, 2003): 135–137.

8. Frank Barnaby, *The Automated Battlefield* (New York: Free Press, 1986); De Landa, *War in the Age of Intelligent Machines*; Edwards, *Closed World*; Mattelart, *Information Society;* Norberg and O'Neill, *Transforming Computer Technology*.

9. De Landa, *War in the Age of Intelligent Machines*.

10. Ibid.

11. Virilio, *War and Cinema*.

12. This was the heralded yet dubious accomplishment of the infamous "smart bombs" used by the U.S. military during the first Gulf War, also known as the "Nintendo war" and the war that "did not take place." Robins, *Into the Image*; Jean Baudrillard, *The Gulf War Did Not Take Place*, trans. Paul Patton (Bloomington: Indiana University Press, 1995). The Gulf War was seen as paradigmatic of a new kind of war, whereby the U.S. military applied advanced imaging, data processing, and weapons technologies in order to dominate the battlefield from afar, incurring almost no losses among its own enlisted soldiers. See also Der Derian, *Virtuous War*.

13. See Foucault, *History of Sexuality*, 135–159; Foucault, *"Society Must Be Defended,"* 239–263.

14. Foucault, *"Society Must Be Defended,"* 244.

15. Ibid., 254.

16. Ibid.

17. Ibid., 256.

18. Ibid., 254.

19. Paul Gilroy, *"There Ain't No Black in the Union Jack": The Cultural Politics of Race and Nation* (London: Hutchinson, 1987): 11.

20. See Majia Nadesan, *Biopower, Governmentality, and Everyday Life* (New York: Routledge, 2008).

21. As Ulrich Beck, Anthony Giddens, Scott Lash, and other theorists of the "risk society" have argued, one of the defining features of the modern period is a change in the nature of risks from fairly predictable dangers largely understood as products of natural events, to more unpredictable sources of hazard and insecurity that are direct or indirect consequences of scientific and technological development. Human development, modernization, and especially globalization have generated manufactured forms of risk that have a particularly unpredictable quality, leading to a spiraling escalation of risk management strategies. Ulrich Beck, *Risk Society: Towards a New Modernity* (London: Sage, 1992); Ulrich Beck, Anthony Giddens, and Scott Lash, *Reflexive Modernization: Politics, Tradition, and Aesthetics in the Modern Social Order* (Palo Alto: Stanford University Press, 1994). Beck has explicitly discussed international terrorism as itself a newly intensified global risk, reflexively generated from political and economic processes of globalization. Beck, "The Silence of Words and Political Dynamics in the World Risk Society," *Logos* 1, no. 4 (2002): 1–18. See also Mark Andrejevic's discussion of "the reflexive threat" in "Interactive (In)security: The Participatory Promise of Ready.gov," *Cultural Studies* 20, nos. 4–5 (2006): 441–458.

22. Lisa Nakamura, "The Socioalgorithmics of Race: Sorting It Out in Jihad Worlds," in *The New Media of Surveillance*, ed. Shoshana Magnet and Kelly Gates (New York: Routledge, 2009): 147.

23. As Sunera Thobani argues, the "war on terror" promotes Western racial solidarity by coding the enemy as non-Western, with a racialized notion of "Muslim" serving as the

quintessential Other of the West. Thobani, *Exalted Subjects: Studies in the Making of Race and Nation* (Toronto: University of Toronto Press, 2007).

24. Sam Keen, *Faces of the Enemy: Reflections of the Hostile Imagination* (San Francisco: Harper & Row, 1986): 12.

25. Karen Engle, "The Face of a Terrorist," *Cultural Studies <=> Critical Methodologies* 7, no. 4 (2007): 401.

26. "Biometric Identifiers and the Modern Face of Terror: New Technologies in the Global War on Terrorism," Hearing Before the Technology, Terrorism, and Government Information Subcommittee of the Judiciary Committee, U.S. Senate, 107th Cong. (November 14, 2001).

27. White House, "President Unveils 'Most Wanted' Terrorists," remarks by the president during announcement at FBI Headquarters, October 10, 2001.

28. Der Derian, *Virtuous War*, 98–121.

29. Paul Virilio, *War and Cinema*, 1.

30. Ibid., 101.

31. Evelyn Hammonds, "New Technologies of Race," in *Processed Lives: Gender and Technology in Everyday Life*, ed. Jennifer Terry and Melodie Calvert (New York: Routledge, 1997): 109.

32. Donna Haraway, *Modest_Witness.*

33. Lauren Berlant, "The Face of America and the State of Emergency," in *Disciplinarity and Dissent in Cultural Studies*, ed. Cary Nelson and Dilip P. Gaonkar (New York: Routledge, 1996): 398. According to Berlant, the "New Face of America" invoked the rhetoric of "the face" found in discourses of social justice, used to solicit personal identification with larger social problems in order to encourage mass sympathy or political commitment (e.g., "the face of AIDS," "the face of welfare," "the face of poverty," etc.). The difficulty with this use of "the face" metaphor, Berlant argues, is its tendency to reduce injustice to something manageable at the level of the individual and to enable "further deferral of considerations that might force structural transformations of public life." Ibid., 435, 406. The use of the "face of terror" trope functioned to produce the opposite effect of personal identification and mass sympathy, instead eliciting mass fear and contempt; however, it likewise encouraged the deferral of structural transformations.

34. Caplan, "This or That Particular Person," 51.

35. Visionics Corporation, *Protecting Civilization from the Faces of Terror: A Primer on the Role Facial Recognition Technology Can Play in Enhancing Airport Security* (2001). The document was originally (but is no longer) available online. For a legal analysis of this document, see David McCormack, "Can Corporate America Secure Our Nation? An Analysis of the Identix Framework for the Regulation and Use of Facial Recognition Technology," *Boston University Journal of Science and Technology Law* 9, no. 1 (2003): 128–155.

36. Visionics Corp., *Protecting Civilization*, 1.

37. Ibid., 5.

38. Sekula, "The Body and the Archive."

39. Ibid., 16

40. Allan Sekula, "Photography Between Labour and Capital," in *Mining Photographs and Other Pictures, 1948–1968: A Selection from the Negative Archives of Shedden Studio, Glace Bay, Cape Breton*, ed. Benjamin H. D. Buchloh and Robert Wilkie (Halifax: Press of the Nova Scotia College of Art and Design, 1983): 194.

41. Lev Manovich, *The Language of New Media* (Cambridge: MIT Press, 2001): 214.
42. Ibid.
43. Poster, "Databases as Discourse," 175–192.
44. Gandy, *Panoptic Sort.*
45. At the time of writing, the *America's Most Wanted* list of the most wanted fugitives could be found at http://www.amw.com/fugitives/most_wanted_lists.cfm. The FBI list of the most wanted terrorists could be found at http://www.fbi.gov/wanted/terrorists/fugitives.htm.
46. Helen Nissenbaum and Lucas D. Introna, "Shaping the Web: Why the Politics of Search Engines Matters," *Information Society* 16, no. 3 (2000): 169–185.
47. Bowker and Star, *Sorting Things Out*, 33, 4.
48. Ibid., 24.
49. http://www.fbi.gov/wanted/terrorists/teralyacoub.htm.
50. Bogard, *Simulation of Surveillance.*
51. Bowker and Star, *Sorting Things Out*, 44.
52. Ibid.
53. The White House had already established the "Terrorist Threat Integration Center," for example.
54. Statement of Donna A. Bucella, director, Terrorist Screening Center, before the National Commission on Terrorist Attacks Upon the United States, January 26, 2004, http://www.9-11commission.gov/hearings/hearing7/witness_bucella.htm.
55. U.S. Department of Justice, Office of the Inspector General, Audit Division, "Review of the Terrorist Screening Center," Audit Report 05-27, June 2005, http://www.justice.gov/oig/reports/FBI/a0527/final.pdf.
56. The report listed many more problems. For example, the screen interface did not display enough information; there were major discrepancies in the size of the database from one phase to another; there were organizational problems at the TSC, including shortage of staff, high turnover, IT management problems, and inexperienced screeners; and so on. U.S. Department of Justice, "Review of the Terrorist Screening Center."
57. Ibid., 52.
58. Bucella statement before the National Commission on Terrorist Attacks Upon the United States, January 26, 2004.
59. Caplan, "This or That Particular Person."
60. Ibid.
61. Foucault, *"Society Must Be Defended,"* 256. See also Agamben, *Homo Sacer.*
62. Virilio, *War and Cinema*; and Virilio, *Vision Machine.*
63. Robins, *Into the Image*, 12.
64. Crary, *Techniques of the Observer*, 2, 1.
65. John Johnston, "Machinic Vision," *Critical Inquiry* 26, no. 1 (1999): 27–48.
66. Ibid., 27.
67. Haraway, "Persistence of Vision," 683.

## NOTES TO CHAPTER 4

1. Neither the VeriFace laptop security software nor the MyFaceID search engine represents the first consumer application of its kind. In 1997, Visionics Corporation released a shrink-wrapped version of its FaceIt facial recognition technology for individual PC

users. (They later sold the rights to a security company called Keyware, although I have not found any evidence that Keyware ever marketed the consumer application of FaceIt.) In 2005, a website called MyHeritage.com had introduced facial recognition software designed by a company called Cognitec into its family genealogy service, encouraging visitors to the site to upload photos of themselves and compare them against a database of photos of famous people. The site also promised that it would eventually help users establish genealogical connections by comparing uploaded photos of their relatives with the photo archives of other site members. Users of MySpace can now use the MyHeritage Celebrity Look-Alikes application to search for celebrity resemblances using their own photos and post the results to their MySpace profile pages.

2. Rose, *Powers of Freedom*, 236–237.

3. Tiziana Terranova has famously examined the provision of "free labor" as a dimension of the cultural economy of advanced capitalist societies. She argues that the Internet provides a special site for eliciting unpaid, voluntary labor, and connects the new unpaid labor of the digital economy to what the Italian autonomists call the "social factory," a shift in the location of work processes from the factory to society. Terranova, "Free Labor: Producing Culture for the Digital Economy," *Social Text 63* 18, no. 2 (2003): 33–57. My argument in this chapter is indebted to hers, but is more closely aligned with Andrejevic's analysis of surveillance and new media interactivity. See Andrejevic, *iSpy*. See also Maurizio Lazzarato, "Immaterial Labour" (1996), http://www.generation-online.org/c/fcimmateriallabour3.htm; and Mark Coté and Jennifer Pybus, "Learning to Immaterial Labour 2.0: MySpace and Social Networks," *Ephemera* 7, no. 1 (2007): 88–106.

4. Winner, "Who Will We Be in Cyberspace?" 64.

5. Michel Foucault, "Technologies of the Self," in *Ethics: Subjectivity and Truth*, ed. Paul Rabinow, trans. Robert Hurley (New York: New Press, 1997): 225.

6. Andrew Barry, *Political Machines: Governing a Technological Society* (New York: Athlone Press, 2001): 31.

7. This summation of neoliberalism draws heavily on Nikolas Rose, "Advanced Liberalism," in *Powers of Freedom*, 137–166. Rose uses the phrase "individualization of security" in "Control," 236.

8. Simon A. Cole and Henry N. Pontell, "'Don't Be a Low Hanging Fruit': Identity Theft as Moral Panic," in *Surveillance and Security: Technological Politics and Power in Everyday Life*, ed. Torin Monahan (New York: Routledge, 2006): 125–148.

9. Ibid. Jennifer Whitson and Kevin Haggerty similarly argue that identity theft protection services "are themselves part of a political strategy whereby institutions are divesting themselves of responsibility for the full social and economic costs of the risks they have produced." Whitson and Haggerty, "Identity Theft and the Care of the Virtual Self," *Economy and Society* 37, no. 4 (2008): 591.

10. See the AuthenTec website, http://www.authentec.com/.

11. Ibid., accessed July 20, 2008.

12. Ibid. The report, "AuthenTec 2008 Consumer Biometrics Survey," is available at http://www.authentec.com/docs/08biometricssurveyresults.pdf.

13. "AuthenTec 2008 Consumer Biometrics Survey," 1.

14. Beyond a note in fine print indicating that the survey was conducted by "the independent online service Zoomerang," there is no discussion of sampling or other statistical methodology—in other words, none of the qualifying information one would expect to

find in an analysis of quantitative survey results. (Zoomerang may in fact be an "independent" firm, but the company website notes that it specializes in "Customer Satisfaction Surveys," suggesting a bias toward a market-friendly approach to research.) The language of the AuthenTec consumer biometrics survey rather easily slips from narrow findings about "respondents" to broader claims about the favorable attitudes of "Americans" or "U.S. consumers" toward the idea of biometric-enabled devices, and the analysis is framed entirely in terms that paint an optimistic picture of the consumer biometrics market.

15. James Hay, "Designing Homes to Be the First Line of Defense: Safe Households, Mobilization, and the New Mobile Privatization," *Cultural Studies* 20, nos. 4–5 (2006): 349–377.

16. Ibid.

17. Jeremy Appel, "High-Tech Security in Low-Income Housing," SecuritySolutions.com (August 1, 1999), http://securitysolutions.com/mag/security_hightech_security_lowincome/.

18. Ibid.

19. Ibid.

20. Ibid.

21. Naomi Klein, *The Shock Doctrine: The Rise of Disaster Capitalism* (New York: Metropolitan, 2007).

22. MyFaceID Facebook platform application information page, http://apps.facebook.com/myfaceid/Default.aspx.

23. Tagg, *Burden of Representation*, 37. As Allan Sekula explains it, the "honorific conventions" of portraiture were "able to proliferate downward." Sekula, "The Body and the Archive," 6.

24. Tagg, *Burden of Representation*, 54.

25. Susan Sontag, *On Photography* (New York: Picador, 1977): 9.

26. Tagg, *Burden of Representation*, 59.

27. Sekula, "The Body and the Archive," 10.

28. In "The Work of Art in the Age of Mechanical Reproduction," Walter Benjamin identifies in the earliest portrait photographs the last remaining vestiges of an "aura," specifically in "the fleeting expression of a human face": "In photography, exhibition value begins to displace cult value [ritual use] all along the line. But cult value does not give way without resistance. It retires into an ultimate retrenchment: the human countenance. It is no accident that the portrait was the focal point of early photography. The cult of remembrance of loved ones, absent or dead, offers a last refuge for the cult value of the picture. For the last time the aura emanates from the early photographs in the fleeting expression of a human face. This is what constitutes their melancholy, incomparable beauty. But as man withdraws from the photographic image, the exhibition value for the first time shows its superiority to the ritual value. To have pinpointed this new state constitutes the incomparable significance of Atget, who, around 1900, took photographs of deserted Paris streets. . . . With Atget, photographs become standard evidence for historical occurrences, and acquire a hidden political significance." Benjamin, "The Work of Art in the Age of Mechanical Reproduction," in *Illuminations*, trans. Hannah Arendt (New York: Schocken, 1968): 225–226.

29. Anthony Giddens, *Modernity and Self-Identity* (Palo Alto: Stanford University Press, 1991): 51.

30. Barry, *Political Machines*.

31. Ibid.,131.

32. Barry argues that interactivity is not a form of discipline in the Foucauldian sense, but a more flexible and active form of agency. Kylie Jarrett has argued, contra Barry, that "interactivity is a technical rendering of neoliberal or advanced liberal power and as such a disciplining technology." Jarret, "Interactivity Is Evil! A Critical Investigation of Web 2.0," *First Monday* 13, no. 3 (2008), http://firstmonday.org/htbin/cgiwrap/bin/ojs/index.php/fm/article/view/2140/1947. My argument here leans more toward Jarrett's claim that interactivity can in fact function as a form of discipline, albeit often in a productive rather than a repressive sense.

33. Andrejevic, *iSpy*, 138.

34. Ibid., 35. The interactive exhibits at science museums have not escaped the logic of interactivity as a consumer monitoring and feedback technique. Barry observes that the intense focus on interactivity in the exhibition planning of science museums in the 1990s led to the marketing-oriented practice of tailoring interactive exhibits to specific audiences: "To an extent unparalleled in the past, the museum visitor became the object of investigation and an element of the museum's internal audit. If interactive technologies are expected to enhance the agency of the visitor and to channel it in the most productive direction, then the specific dynamic of this agency must itself be known." Barry, *Political Machines*, 142.

35. Ritendra Datta, Dhiraj Joshi, Jia Li, and James Z. Wang, "Image Retrieval: Ideas, Influences, and Trends of the New Age," *ACM Computing Surveys* 40, no. 2 (2008): 1–60.

36. Ibid., 14.

37. Google Image Search, http://images.google.com/, accessed July 29, 2008.

38. Google Image Labeler, http://images.google.com/imagelabeler/, accessed July 29, 2008.

39. http://www.polarrose.com/, accessed August 2, 2008.

40. The invitation for user experimentation and feedback is another strategy popularized by Google: featuring the "beta" versions of new online software programs so users can experiment with technologies that "aren't quite ready for primetime." http://labs.google.com/. At the "Google Labs" page, the company lists applications still in development and invites visitors to "please play with these prototypes and send your comments directly to the Googlers who developed them." Ibid.

41. MyFaceID Facebook platform application information page, http://apps.facebook.com/myfaceid/.

42. Ibid.

43. Technology testing, such as the Facial Recognition Vendor Test sponsored by the U.S. National Institute of Standards and Technology, has compared human accuracy rates at matching facial image pairs with the performance of computer algorithms. According to the FRVT 2006 study, the best performing algorithms were more accurate than humans at correctly matching faces. See P. Jonathan Phillips, W. Todd Scruggs, Alice J. O'Toole, Patrick J. Flynn, Kevin W. Bowyer, Cathy L. Schott, and Matthew Sharpe, *FRVT 2006 and ICE 2006 Large-Scale Results* (Gaithersburb, MD: National Institute of Standards and Technology, 2007), iris.nist.gov/ice/FRVT2006andICE2006LargeScaleReport.pdf. In a slightly different evaluative context, Datta and others note that the collection of manual photo tagging data "has the dual advantage of: (1) facilitating text-based querying, and (2) building reliable training datasets for content-based analysis and automatic annotation algorithms." Datta et al., "Image Retrieval," 12.

44. Both government agencies and local police forces have admitted using social networking sites to investigate individuals and organizations. According to Andrejevic, "The Pentagon's National Security Agency has already funded research into 'the mass harvesting of the information people post about themselves on social networks.' . . . Posting information on MySpace, in other words, can double as a form of active participation in creating a government data dossier about oneself." Andrejevic, *iSpy*, 176.

45. Mark Andrejevic, "The Work of Watching One Another: Lateral Surveillance, Risk, and Governance," *Surveillance and Society* 2, no. 4 (2005): 479–497.

46. An essay in the *Harvard Law Review* argues that Congress should "require facial recognition search engines to provide an opt-out regime that allows users to remove their names from these search engines' databases." Anonymous, "2006 Oliver Wendell Holmes Lecture: In the Face of Danger: Facial Recognition and the Limits of Privacy Law," *Harvard Law Review* 120, no. 7 (2007): 1872.

47. Hille Koskela, "Webcams, TV Shows, and Mobile Phones: Empowering Exhibitionism," *Surveillance and Society* 2, nos. 2–3 (2004): 199–215.

48. Peters, *Speaking into the Air*, 224.

## NOTES TO CHAPTER 5

1. Joseph Weizenbaum, *Computer Power and Human Reason* (New York: Freeman, 1976).

2. Colby et al. quoted in ibid., 5.

3. Weizenbaum, *Computer Power and Human Reason*, 5–6.

4. See Janet Murray, *Hamlet on the Holodeck: The Future of Narrative in Cyberspace* (Cambridge: MIT Press, 1997): 68–77, 214–247; and Sherry Turkle, *Life on the Screen: Identity in the Age of the Internet* (New York: Touchstone, 1995): 108–114.

5. The term "blank somatic surface" is from Sobieszek, *Ghost in the Shell*, 86.

6. There is no agreed-on acronym for either facial recognition technology or automated facial expression analysis. Ying-Li Tian, Takeo Kanade, and Jeffrey Cohn use the acronym AFEA for the latter. Tian, Kanade, and Cohn, "Facial Expression Analysis," in *Handbook of Face Recognition*, ed. Stan Z. Li and Anil K. Jain (New York: Springer, 2004): 247–276.

7. Otniel E. Dror, "The Scientific Image of Emotion: Experience and Technologies of Inscription," *Configurations* 7, no. 3 (1999): 367.

8. Peters, *Speaking into the Air*, 228.

9. Jeffrey F. Cohn and Takeo Kanade "Use of Automated Facial Image Analysis for Measurement of Emotion Expression," in *Handbook of Emotion Elicitation and Assessment*, ed. James A. Cohn and John J. B. Allen (New York: Oxford University Press, 2007): 232.

10. Beat Fasel and Juergen Luettin, "Automatic Facial Expression Analysis: A Survey," *Pattern Recognition* 36, no. 1 (2003): 271.

11. Tian, Kanade, and Cohn, "Facial Expression Analysis," 247. Another application frequently mentioned by those interested in applications of automated expression analysis involves its potential use for helping children with autism develop better face processing skills. For example, among the experimental applications of the Computer Expression Recongition Toolbox (CERT), developed at the Machine Perception Laboratory at the University of California, San Diego, is "a project to give feedback on facial expression production to children with autism." Marian Bartlett, Gwen Littlewort, Tingfan Wu, and Javier Movellan, "Computer Expression Recognition Toolbox," Eighth International IEEE

Conference on Automatic Face and Gesture Recognition, Amsterdam, The Netherlands, September 17–19, 2008, http://mplab.ucsd.edu/~marni/pubs/Bartlett_demo_FG2008.pdf. See also the website of the "Let's Face It!" project, "a multimedia, computer-based intervention that is designed to teach face processing skills to children with autism," http://web.uvic.ca/~letsface/letsfaceit/index.php. I do not spend time on the autism applications of AFEA in this chapter because, while developing technologies to help children with autism is clearly an important and worthwhile endeavor, it is not likely to be the most prevalent or significant use of AFEA, nor is it likely to reach more than a narrow selection of children with autism. I choose instead to focus on applications that are likely to find more widespread social uses and therefore potentially have more profound and pervasive effects on the meanings and experiences of affect.

12. Ibid.

13. Associated Press, "Disney to Create Laboratory to Test High-Tech Advertising for ABC, ESPN," *International Herald Tribune*, May 13, 2008.

14. One exception is Lisetti and Schiano, "Automatic Facial Expression Interpretation," 222–227. For a discussion of the ethical implications of affective computing applications more generally, see Rosalind Picard, "Potential Concerns," in *Affective Computing* (Cambridge: MIT Press, 1998): 113–138.

15. Tian, Kanade, and Cohn, "Facial Expression Analysis," 247.

16. See also Winner, "Who Will We Be in Cyberspace?" 64.

17. Nikolas Rose, *Inventing Our Selves: Psychology, Power, and Personhood* (New York: Cambridge University Press, 1998): 88–89.

18. As Alan Fridlund explains, the science of phrenology, associated with Franz Joseph Gall, came to overshadow the physiognomy of Lavater in Europe, but "despite Gall's antipathy for Lavaterian physiognomy and his desire to keep phrenology separate, the two were often practiced together." Fridlund, *Human Facial Expression: An Evolutionary View* (San Diego: Academic Press, 1994): 5.

19. John Welchman, "Face(t)s: Notes on Faciality," *ArtForum* (November 1988): 133.

20. Peter Hamilton and Roger Hargreaves, *The Beautiful and the Damned: The Creation of Identity in Nineteenth-Century Photography* (London: Lund Humphries, 2001): 65.

21. Graeme Tytler, *Physiognomy in the European Novel: Faces and Fortunes* (Princeton: Princeton University Press, 1982): 78.

22. Sander Gilman, *The Face of Madness: Hugh W. Diamond and the Origin of Psychiatric Photography* (New York: Brunner/Mazel, 1976): 9.

23. Hamilton and Hargreaves, *The Beautiful and the Damned*, 65.

24. Gunning, "In Your Face," 6.

25. Ibid., 77.

26. Fridlund, *Human Facial Expression.*

27. Quoted in Hamilton and Hargreaves, *The Beautiful and the Damned*, 77.

28. Ibid.

29. Gunning, "In Your Face."

30. Béla Balázs, quoted in ibid., 1.

31. As Sobieszek explains in *Ghost in the Shell*, even as the West transitioned from the visual "physiognomic culture" of the eighteenth and nineteenth centuries to the linguistic "psychoanalytic culture" of the early twentieth, there remained an overwhelming preoccupation with looking beneath the surface to excavate the depths of the individual's soul.

32. Jonathan Sterne, *The Audible Past: Cultural Origins of Sound Reproduction* (Durham: Duke University Press, 2003): 10.

33. Ibid.

34. Norbert Elias, *The Civilizing Process*, rev. ed., trans. Edmund Jephcott (Oxford: Blackwell, 1994).

35. Ibid., 450.

36. As Elias writes, "At present we are so accustomed to the existence of these more stable monopolies of force and the greater predictability of violence resulting from them, that we scarcely see their importance for the structure of our conduct and our personality. We scarcely realize how quickly what we call our 'reason,' this relatively farsighted and differentiated steering of our conduct, with its high degree of affect-control, would crumble or collapse if the anxiety-inducing tensions within and around us changed, if the fears affecting our lives suddenly became much stronger or much weaker or, as in many simpler societies, both at once, now stronger, now weaker." Elias, *Civilizing Process*, 441.

37. Ibid.

38. Ibid., 443.

39. Ibid.

40. Dror, "Scientific Image of Emotion," 364.

41. Ibid., 391.

42. Ibid., 393–394.

43. Ibid., 394.

44. Hacking, "Making Up People," 229.

45. Ibid., 236.

46. Ibid., 235.

47. Rose, *Inventing Our Selves*, 10.

48. Erika L. Rosenberg, "Introduction: The Study of Spontaneous Facial Expressions in Psychology," in *What the Face Reveals: Basic and Applied Studies of Spontaneous Expression Using the Facial Action Coding System*, ed. Paul Ekman and Erika L. Rosenberg (New York: Oxford University Press, 1997): 10.

49. Jill Bolte Taylor, "The 2009 *Time* 100: Paul Ekman," *Time* (April 30, 2009).

50. Despite the high profile of Ekman's theories and his prominence as a preeminent facial expression expert, the relationship between facial expressions and emotions is a controversial issue and remains open to debate among psychologists. The psychologist Alan Fridlund divides the psychological literature on facial expression between the "Emotion View," represented by Ekman's theories of facial affect, and the "Behavioral Ecology View," held by Fridlund and others who maintain that facial displays have no necessary relationship to emotions but instead are entirely context-specific social signals of intent. Fridlund, *Human Facial Expression.* The Behavioral Ecology View should not be confused with social constructivism, however. While it holds that "there are neither fundamental emotions nor fundamental expressions of them," this view nevertheless has major Darwinian underpinnings: "As manipulations serving [social] intent," writes Fridlund, "[facial] displays are issued when they will optimally enhance cultural or genetic inclusive fitness." Ibid., 139. Consistent with sociobiology, the Behavioral Ecology view maintains that facial displays are not learned but instead are mediated by an *innate* form of social cognition, and that displays have evolved according to the pressures of natural selection, including the responses of others to those displays. Both the Emotion

and Behavioral Ecology Views of facial expressions take an evolutionary perspective and posit facial expressions as innate forms of human behavior. Still, the differences between these theories suggest that the relationship between facial expressions and the internal state of the person is by no means a settled question.
James M. Caroll and James A. Russell have argued that facial expressions do not necessarily bear any direct relationship to emotions, and emphasize the importance of context in making sense of the intentions behind facial expressions and their reception. See "Do Facial Expressions Signal Specific Emotions? Judging Emotion from the Face in Context," *Journal of Personality and Social Psychology* 70, no. 2 (1996): 105–218.

51. Paul Ekman and Wallace Friesen, "Measuring Facial Movement with the Facial Action Coding System," in *Emotion in the Human Face*, ed. Paul Ekman (Cambridge: Cambridge University Press, 1982): 179.

52. Motoi Suwa, Noboru Sugie, and Keisuke Fujimura, "A Preliminary Note on Pattern Recognition of Human Emotional Expression," in *Proceedings of the 4th International Joint Conference on Pattern Recognition, November 7–10, 1978, Kyoto, Japan* (New York: Institute of Electrical and Electronics Engineers, 1979): 408–410.

53. Ibid., 408.

54. Ibid., 410.

55. Tian, Kanade, and Cohn, "Facial Expression Analysis."

56. Ibid., 254.

57. According to Elias, "The behaviour patterns of our society, imprinted on individuals from early childhood as a kind of second nature and kept alert in them by a powerful and increasingly strictly organized social control, are to be explained, it has been shown, not in terms of general, ahistorical human purposes, but as something which has evolved from the totality of Western history, from the specific forms of behaviour that developed in its course and the forces of integration which transformed and propagated them." Elias, *Civilizing Process*, 441. See also Erving Goffman's work on "face-work," *Interaction Ritual: Essays in Face-to-Face Behavior* (New Brunswick, NJ: Aldine Transaction, 1967/2005).

58. Goffman, *Interaction Ritual*, 5.

59. Ibid., 10.

60. Hubert Dreyfus, *What Computers Still Can't Do* (Cambridge: MIT Press, 1972/1992).

61. Ibid., 103.

62. Ibid., 109.

63. Lisetti and Schiano offer a clear explanation of neural networks: "Connectionist models are networks of simple interconnected processing units that operate in parallel. The term neural network is also sometimes used to refer to the same approach because the processing units of the network are neuron-like, and they function metaphorically as neurons. Each unit receives inputs from other units, sums them, and calculates the output to be sent to the other units that this unit is connected to. Learning or the acquiring of knowledge results from the modification of the strength of the connection (the *weight*, akin to a synaptic connection) between interconnected units: the higher the weight, the stronger the connection, the higher the propensity of one neuron to cause its neighboring neurons to become active or to fire, the more learning happens, i.e., the more knowledge is retained in the strengths of the connection. Neural networks can learn patterns for object recognition by *unsupervised clustering* (i.e., no specific target is provided as the target label/response to be learned), or by *supervised learning/training by example* (i.e., the

input and the corresponding desired target response are both given). Networks learn by successively modifying the strengths of the connections between units, aiming at reducing the error at the output." Lisetti and Schiano, "Automatic Facial Expression Interpretation," 203.

64. Dreyfus, *What Computers Still Can't Do*, xxxvii.

65. Ibid., xxxviii.

66. Jonathan Sterne, "Bourdieu, Technique, and Technology," *Cultural Studies* 17, no. 3 (2003): 377.

67. Fasel and Luettin, "Automatic Facial Expression Analysis," 260.

68. Ibid., 260.

69. Tian, Kanade, and Cohn, "Facial Expression Analysis," 249.

70. Joseph C. Hager and Paul Ekman, "Essential Behavior Science of the Face and Gesture that Computer Scientists Need to Know," http://www.face-and-emotion.com/dataface/misctext/iwafgr.html.

71. In an essay posted online, Joseph Hager and Paul Ekman explain the Shannon-Weaver mathematical model of communication as the basic principle underlying the development of FACS. Hager and Ekman, "Essential Behavior Science." Hager was a student of Ekman's and is now an entrepreneur whose business combines computing and psychology. Ekman hired Hager in 1991 to manage his National Science Foundation research project for automating the Facial Action Coding System. Hager's biography is available online at http://www.joseph.hager.name/.

72. Ekman and Friesen, "Measuring facial movement with the Facial Action Coding System," 180.

73. Ibid., 181.

74. Marian Stewart Bartlett, Joseph C. Hager, Paul Ekman, and Terrence Sejnowski, "Measuring Facial Expressions by Computer Image Analysis," *Psychophysiology* 36, no. 2 (1999): 254.

75. Rosenberg, "Introduction," 15.

76. Cohn and Kanade, "Use of Automated Facial Image Analysis for Measurement of Emotion Expression," 222.

77. Ibid., 254.

78. Bartlett et al., "Measuring Facial Expressions by Computer Image Analysis," 253.

79. Ibid.

80. Ekman and Friesen, "Measuring Facial Movement with the Facial Action Coding System," 195.

81. Ekman and Friesen conducted tests to measure the reliability of FACS—the degree to which different coders agreed on the coding of particular faces. They found relatively high intercoder agreement: most of the coders saw the same facial actions on the sample of images used for the study, with only a minimal amount of disagreement. (There was slightly less agreement among coders about the intensity of facial actions.)

82. Jeffrey Cohn, "Foundations of Human Centered Computing: Facial Expression and Emotion," *Proceedings of the 8th International Conference on Multimodal Interfaces* (New York: Association for Computing Machinery, 2006): 235.

83. N. Katherine Hayles, *How We Became Posthuman: Virtual Bodies in Cybernetics, Literature, and Informatics* (Chicago: University of Chicago Press, 1999): 51–57.

84. Ibid., 54.

85. Ekman and Friesen, "Measuring Facial Movement with the Facial Action Coding System," 209.

86. Joseph C. Hager, "History of FACSAID," http://www.face-and-emotion.com/dataface/facsaid/history.jsp.

87. Rosenberg, "Introduction," 15.

88. Joseph C. Hager, "Facial Action Coding System Affect Interpretation Dictionary (FACSAID)," http://www.face-and-emotion.com/dataface/facsaid/description.jsp.

89. Hager, "History of FACSAID."

90. See Hager and Ekman, "Essential Behavioral Science of the Face and Gesture." See also Paul Ekman, "Strong Evidence for Universals in Facial Expressions: A Reply to Russell's Mistaken Critique," *Psychological Bulletin* 115, no. 2 (1994): 268–287.

91. Bowker and Star, *Sorting Things Out*, 47. Bowker and Star use the term "reverse engineering" to explain their own analytical approach rather than the work that classification systems do.

92. Armand Mattelart, *Theories of Communication* (Thousand Oaks, CA: Sage, 1998): 45.

93. Ekman and Friesen, "Measuring Facial Movement with the Facial Action Coding System," 180, 185–186.

94. Another effort to establish FACS as the basis for the discovery of the *facts* of facial expression can be found in the phonetic similarity between the word "facts" and the acronym "FACS" (they are phonetically almost identical). The acronym represents a claim to truth about the coding system, a direct attempt at rhetorical closure, and a move aimed at selling FACS as an established "ground truth."

95. Daston and Galison, "Image of Objectivity," 82.

96. Ibid., 120.

97. As Nikolas Rose argues, psychology's applications cannot be separated out from the discipline as such: "Psychology was only able to differentiate from medicine, philosophy, and physiology, only able to 'disciplinize' itself, on the basis of its social vocation, its elaboration within the educational, penal, military, and industrial apparatuses." Rose, *Inventing Our Selves*, 82.

98. Ibid., 81.

99. Ibid.

100. Ibid.

101. Nigel Thrift, *Non-representational Theory: Space, Politics, Affect* (New York: Routledge, 2008).

102. Ibid., 182.

103. Ibid., 183.

104. Ibid., 194.

105. Ibid., 187.

106. Ibid.

107. Ibid.

108. Ibid., 185.

109. Rose, *Inventing Our Selves*, 103.

110. Ibid., 87.

111. Dacher Keltner, Terrie E. Moffit, and Magda Stouthamer-Leober, "Facial Expressions of Emotion and Psychopathology in Adolescent Boys," *What the Face Reveals: Basic and Applied Studies of Spontaneous Expression Using the Facial Action Coding System*, ed. Paul Ekman and Erika L. Rosenberg (New York: Oxford University Press, 1997): 440.

112. Ibid.

113. Ibid., 437.

114. Ibid., 444. The researchers were principally concerned with antisocial behavior and child development, and they acknowledged the fundamental, albeit micro-level social basis of facial expressions: "We believe that the responses that facial expressions evoke in others mediate the social relationships that externalizing, internalizing, and nondisordered adolescents develop." Ibid., 445–446.

115. Rose, *Inventing Our Selves*, 11.

116. Toby Miller, *Makeover Nation: The United States of Reinvention* (Columbus: Ohio State University Press, 2008): 39.

117. John O'Shaughnessy and Nicholas Jackson O'Shaughnessy, *The Marketing Power of Emotion* (Oxford: Oxford University Press, 2003).

118. Flemming Hansen and Sverre Riis Christensen, *Emotions, Advertising, and Consumer Choice* (Copenhagen, Denmark: Copenhagen Business School Press, 2007): 154.

119. Miller, *Makeover Nation*, 2; Robert McChesney and John Bellamy Foster, quoted in ibid., 36.

120. Ekman, *Telling Lies*.

121. Ibid., 127.

122. Paul Ekman, "How to Spot a Terrorist on the Fly," *Washington Post*, October 29, 2006.

123. See Jeffrey F. Cohn, Takeo Kanade, Tsuyoshi Moriyama, Zara Ambadar, Jing Xiao, Jiang Gao, and Hiroki Imamura, "Final Report, CIA Contract #2000-A128400-000: A Comparative Study of Alternative FACS Coding Algorithms" (November 9, 2001), http://mplab.ucsd.edu/grants/project1/publications/pdfs/CMUFinalReport.pdf.

124. Ekman, *Telling Lies*, 129.

125. Marian Bartlett Stewart, *Face Image Analysis by Unsupervised Learning* (Boston: Kluwer, 2001): 79–80.

126. Lisetti and Schiano, "Automatic Facial Expression Interpretation," 197.

127. Melissa Littlefield, "Constructing the Organ of Deceit: The Rhetoric of fMRI and Brain Fingerprinting in Post-9/11 America," *Science, Technology, and Human Values* 34, no. 3 (2009): 380.

128. *Frye v. United States* (1923) 54 App. D. C. 46, 293 F. 1013.

129. William C. Iacono, "Forensic 'Lie Detection': Procedures Without Scientific Basis," *Journal of Forensic Psychology Practice* 1, no. 1 (2001): 75.

130. This was the finding of a study by the psychologist David Lykken, cited in Ken Adler, *The Lie Detectors: The History of an American Obsession* (New York: Free Press, 2007): xiv.

131. Ken Adler, *Lie Detectors*, xiv.

132. Littlefield, "Constructing the Organ of Deceit," 366.

133. Ibid., 371.

134. Cultural representations of deception detection technologies serve an especially important ideological function in this regard. This is perhaps best illustrated on the fictional television show inspired by Ekman's work, *Lie to Me*, where deception detection is presented as an absolute science. In the show, the practice of spotting lies is a professional and decidedly *human* skill, albeit one often assisted by visual media technologies. But the fact that the lead character, Dr. Cal Lightman, can identify deception with absolute certainty in every

case disavows fundamental uncertainties about techniques for delineating truth and deception, as well as the ambiguities that permeate their discursive and behavioral manifestations.

135. Kirsten Boehner and her colleagues likewise argue that "emotion is an intersubjective phenomenon, arising in encounters between individuals or between people and society, an aspect of the socially-organized lifeworld we both inhabit and reproduce." Kirsten Boehner, Rogerio DePaula, Paul Dourish, and Phoebe Sengers, "How Emotion Is Made and Measured," *International Journal of Human-Computer Studies* 65, no. 4 (2007): 280.

136. Cohn, "Foundations of Human Centered Computing," 5.

137. Lisetti and Schiano, "Automatic Facial Expression Interpretation," 186.

138. Ibid.

139. Ibid., 199.

140. The turn to "affective computing" corresponds with the emergence of theories in neuroscience and psychology about the relationship between cognition and emotion, as well as recognition that intelligence and rational thought have emotional dimensions. Boehner et al., "How Emotion Is Made and Measured," 277; Lisetti and Schiano, "Automatic Facial Expression Interpretation," 208–211. In neuroscience, see Antonio Damasio, *Descartes' Error: Emotion, Reason and the Human Brain* (New York: Quill, 1995).

141. Jay David Bolter and Richard Grusin, *Remediation: Understanding New Media* (Cambridge: MIT Press, 2000): 11.

142. Winner, "Who Will We Be in Cyberspace?" 68.

143. Lisetti and Schiano, "Automatic Facial Expression Interpretation," 187.

144. Ibid., 199.

145. Boehner et al., "How Emotion Is Made and Measured," 275–291.

146. Ibid., 280.

147. Ibid., 284.

148. David Golumbia, *The Cultural Logic of Computation* (Cambridge: Harvard University Press, 2009): 5.

149. Emmanuel Levinas, *Totality and Infinity: An Essay on Exteriority*, trans. Alphonso Lingis (Pittsburgh: Duquesne University Press, 1969).

150. Lyon, *Surveillance Society.*

151. On "co-presence," see also Anthony Giddens's discussion of Erving Goffman in *The Constitution of Society: Outline of the Theory of Structuration* (Berkeley: University of California Press, 1984): 36, 69–72.

152. Peters, *Speaking into the Air*, 9.

153. Ibid.

154. Ibid.

155. Hayles, *How We Became Posthuman*, 54.

156. Weizenbaum, *Computing Power and Human Reason*, 9.

157. Bruno Latour, "Mixing Humans and Non-humans Together: The Sociology of a Door-Closer," *Social Problems* 35, no. 3 (1988): 301.

## NOTES TO THE CONCLUSION

1. The image that circulated in the press was actually an average of multiple video frames. Julia Scheeres, "Video Forensics: Grainy to Guilty," *Wired* (January 30, 2002), http://www.wired.com/politics/law/news/2002/01/50036.

2. Vicki Bruce and Andy Young, *In the Eye of the Beholder: The Science of Face Perception* (Oxford: Oxford University Press, 1998): 3.

3. Ibid.

4. Ibid.

5. Stan Z. Li and Anil K. Jain, "Introduction," in *Handbook of Face Recognition*, ed. Stan Z. Li and Anil K. Jain (New York: Springer, 2005): 1.

6. Welchman, "Face(t)s," 131.

7. Levinas, *Totality and Infinity*.

8. Peters, *Speaking into the Air*, 9.

9. The transcript of the October 2, 2008, vice presidential debate is available online at http://www.cnn.com/2008/POLITICS/10/02/debate.transcript/.

10. Ibid.

11. Harvey, *Brief History of Neoliberalism*, 5.

12. Miller, *Makeover Nation*, 3.

13. Ibid.

14. The term "surveillant assemblage" is from Haggerty and Ericson, "Surveillant Assemblage."

15. See Mark Andrejevic, "The Kinder, Gentler Gaze of Big Brother: Reality TV in the Era of Digital Capitalism," *New Media and Society* 4, no. 2 (2002): 251–270. Andrejevic does not argue that the contemporary market-oriented version of Big Brother is in fact kinder and gentler than Orwell's vision, but that reality TV portrays him as such, encouraging the audience to view being watched as innocuous, pleasurable, and even a form of self-expression and self-actualization. The term "panoptic sort" is from Oscar Gandy, *Panoptic Sort*.

16. The Visionics CEO Joseph Atick used the terms "innocent majority" and "honest majority" repeatedly in his post-9/11 interviews with the press. See, for example, "Joseph Atick: How the Facial Recognition Security System Works," CNN.com/Community (October 1, 2001), http://archives.cnn.com/2001/COMMUNITY/10/01/atick/.

17. For discussions of issues of security and belonging in the discourse, policy, and practice of international relations, see Keith Krause and Michael C. Williams, eds., *Critical Security Studies* (Minneapolis: University of Minnesota Press, 1997). For example, Simon Dalby in this volume discusses the attempts in the post–Cold War context of the 1990s to reformulate security policies around broader notions of "cooperative security" and global human rights, and more willingness in the security community to acknowledge that the political order itself may generate insecurity. Dalby argues that "security" is an "essentially contested concept," and that from a critical security studies perspective, "states and alliances are not the assumed starting point for analysis of state behavior and the application of strategic planning, but the political problem to be analyzed." Dalby, "Contesting an Essential Concept: Reading the Dilemmas in Contemporary Security Discourse," in Krause and Williams, *Critical Security Studies*, 9. For more recent discussions of these issues in the post-9/11 context, see Muller, "(Dis)qualified Bodies"; Rygiel, "Protecting and Proving Identity"; and Salter, "When the Exception Becomes the Rule."

18. Here I am taking seriously Irma van der Ploeg's argument that a reified conceptualization of biometrics, while perhaps better suited to raising public alarm about the technologies, leaves little room for intervening on the design of technical systems. Van der Ploeg, "Biometrics and Privacy," 100.

19. Introna, "*Dis*closive Ethics." See also Introna and Wood, "Picturing Algorithmic Surveillance."

20. Introna, "*Dis*closive Ethics," 77.

21. Judith Donath, "Mediated Faces," in *Proceedings of the 4th International Conference on Cognitive Technology: Instruments of the Mind*, ed. Meurig Beynon, Chrystopher L. Nehaniv, and Kerstin Dautenhahn (London: Springer-Verlag, 2001): 373.

22. Ibid., 374.

23. Golumbia, *Cultural Logic of Computation*; Hayles, *How We Became Posthuman*; Vincent Mosco, *The Digital Sublime: Myth, Power, and Cyberspace* (Cambridge: MIT, 2004); Robins, *Into the Image*; Weizenbaum, *Computer Power and Human Reason*.

24. Nicholas Negroponte, *Being Digital* (New York: Vintage, 1995): 129.

25. In an explication of Deleuze's concept of faciality, Richard Rushton notes that "the face does not conform to the cause-effect model of communication. . . . The face is a link between destination and an origin; the face arrives from somewhere and is on its way to somewhere else." Rushton, "Response to Mark B. N. Hansen's Affect as Medium, or the 'Digital-Facial-Image,'" *Journal of Visual Culture* 3, no. 3 (2004): 225. See also Rushton, "What Can a Face Do? On Deleuze and Faces," *Cultural Critique* 51 (2002): 219–237.

# Bibliography

Adler, Ken. *The Lie Detectors: The History of an American Obsession*. New York: Free Press, 2007.

Agamben, Giorgio. *Homo Sacer: Sovereign Power and Bare Life*, translated by Daniel Heller-Roazen. Palo Alto: Stanford University Press, 1998.

Agre, Phil. "Your Face Is Not a Bar Code: Arguments Against Automatic Face Recognition in Public Places" (2001). http://polaris.gseis.ucla.edu/pagre/bar-code.html.

Andreas, Peter. "Introduction: The Wall After the Wall." In *The Wall Around the West: State Borders and Immigration Controls in North America and Europe*, edited by Peter Andreas and Timothy Snyder, 1–11. Lanham, MD: Rowman and Littlefield, 2000.

Andrejevic, Mark. "Interactive (In)security: The Participatory Promise of Ready.gov." *Cultural Studies* 20, nos. 4–5 (2006): 441–458.

———. *iSpy: Surveillance and Power in the Interactive Era*. Lawrence: University Press of Kansas, 2007.

———. "The Kinder, Gentler Gaze of Big Brother: Reality TV in the Era of Digital Capitalism." *New Media and Society* 4, no. 2 (2002): 251–270.

———. "The Work of Being Watched: Interactive Media and the Exploitation of Self-Disclosure." *Critical Studies in Media Communication* 19, no. 2 (2002): 230–248.

———. "The Work of Watching One Another: Lateral Surveillance, Risk, and Governance." *Surveillance and Society* 2, no. 4 (2005): 479–497.

Anonymous. "2006 Oliver Wendell Holmes Lecture: In the Face of Danger: Facial Recognition and the Limits of Privacy Law." *Harvard Law Review* 120, no. 7 (2007): 1872–1891.

Ball, Kirstie, and Frank Webster. "The Intensification of Surveillance." In *The Intensification of Surveillance: Crime, Terrorism, and Warfare in the Information Age*, edited by Kirstie Ball and Frank Webster, 1–15. London: Pluto Press, 2003.

Ballantyne, Michael, Robert S. Boyer, and Larry Hines, "Woody Bledsoe: His Life and Legacy." *AI Magazine* 17, no. 1 (1996): 7–20.

Barnaby, Frank. *The Automated Battlefield*. New York: Free Press, 1986.

Barry, Andrew. *Political Machines: Governing a Technological Society*. New York: Athlone Press, 2001.

Barthes, Roland. *Camera Lucida: Reflections on Photography*, translated by Richard Howard. New York: Hill and Wang, 1981.

Bartlett, Marian Stewart. *Face Image Analysis by Unsupervised Learning*. Boston: Kluwer, 2001.

Bartlett, Marian Stewart, Joseph C. Hager, Paul Ekman, and Terrence Sejnowski. "Measuring Facial Expressions by Computer Image Analysis." *Psychophysiology* 36, no. 2 (1999): 253–263.

Baudrillard, Jean. *The Gulf War Did Not Take Place*, translated by Paul Patton. Bloomington: Indiana University Press, 1995.
Beck, Ulrich. *Risk Society: Towards a New Modernity*. London: Sage, 1992.
———. "The Silence of Words and Political Dynamics in the World Risk Society." *Logos* 1, no. 4 (2002): 1–18.
Beck, Ulrich, Anthony Giddens, and Scott Lash. *Reflexive Modernization: Politics, Tradition, and Aesthetics in the Modern Social Order*. Palo Alto: Stanford University Press, 1994.
Beniger, James R. *The Control Revolution: Technological and Economic Origins of the Information Society*. Cambridge: Harvard University Press, 1986.
Benjamin, Walter. "The Work of Art in the Age of Mechanical Reproduction." In *Illuminations*, translated by Hannah Arendt, 225–226. New York: Schocken, 1968.
Berger, John. *Ways of Seeing*. London: Penguin, 1972.
Berlant, Lauren. "The Face of America and the State of Emergency." In *Disciplinarity and Dissent in Cultural Studies*, edited by Cary Nelson and Dilip P. Gaonkar, 397–440. New York: Routledge, 1996.
Biernoff, Suzannah. "Carnal Relations: Embodied Sight in Merleau-Ponty, Roger Bacon, and St Francis." *Journal of Visual Culture* 4, no. 1 (2005): 39–52.
Blackburn, Duane M., Mike Bone, and P. Jonathan Phillips. *Facial Recognition Vendor Test 2000: Executive Overview* (December 2000). http://www.frvt.org/DLs/FRVT2000_Executive_Overview.pdf.
Boehner, Kirsten, Rogerio DePaula, Paul Dourish, and Phoebe Sengers. "How Emotion Is Made and Measured." *International Journal of Human-Computer Studies* 65, no. 4 (2007): 275–291.
Bogard, William. *The Simulation of Surveillance: Hypercontrol in Telematic Societies*. New York: Cambridge University Press, 1996.
Bolter, Jay David, and Richard Grusin. *Remediation: Understanding New Media*. Cambridge: MIT Press, 2000.
Bowker, Geoffrey C., and Susan Leigh Star. *Sorting Things Out: Classification and Its Consequences*. Cambridge: MIT Press, 2000.
Boyle, James. "The Second Enclosure Movement and the Construction of the Public Domain." *Law and Contemporary Problems* 66 (Winter–Spring 2003): 33–74.
———. *Shamans, Software, and Spleens: Law and the Construction of the Information Society*. Cambridge: Harvard University Press, 1996.
Bruce, Vicki, Peter J. B. Hancock, and A. Mike Burton. "Human Face Perception and Identification." In *Face Recognition: From Theory to Applications*, edited by Harry Wechsler et al., 51–72. New York: Springer, 1998.
Bruce, Vicki, and Andy Young. *In the Eye of the Beholder: The Science of Face Perception*. Oxford: Oxford University Press, 1998.
Burnham, David. *The Rise of the Computer State*. New York: Random House, 1983.
Caplan, Jane. "'This or That Particular Person': Protocols of Identification in Nineteenth-Century Europe." In *Documenting Individual Identity: The Development of State Practices in the Modern World*, edited by Jane Caplan and John Torpey, 49–66. Princeton: Princeton University Press, 2001.
Caplan, Jane, and John Torpey, eds. *Documenting Individual Identity: The Development of State Practices in the Modern World*. Princeton: Princeton University Press, 2001.

Carlson, Matthew. "Tapping into TiVo: Digital Video Recorders and the Transition from Schedules to Surveillance in Television." *New Media and Society* 8, no. 1 (2006): 97–115.
Carroll, James M., and James A. Russell. "Do Facial Expressions Signal Specific Emotions? Judging Emotion from the Face in Context." *Journal of Personality and Social Psychology* 70, no. 2 (1996): 105–218.
Cartwright, Lisa. "Photographs of 'Waiting Children': The Transnational Adoption Market." *Social Text 74* 21, no. 1 (2003): 83–108.
Castells, Manuel. *The Information Age: Economy, Society and Culture*, volume 3: *End of the Millennium*. Malden, MA: Blackwell, 1998.
Chellappa, Rama, Charles L. Wilson, and Saad Sirohey. "Human and Machine Recognition of Faces: A Survey." *Proceedings of the IEEE* 83, no. 5 (1995): 710–711.
Christie, Nils. *Crime Control as Industry: Towards Gulags, Western Style*. 3rd ed. New York: Routledge, 2000.
Christopherson, Susan. "The Fortress City: Privatized Spaces, Consumer Citizenship." In *Post-Fordism: A Reader*, edited by Ash Amin, 409–426. Cambridge: Blackwell, 1994.
Cohn, Jeffrey. "Foundations of Human Centered Computing: Facial Expression and Emotion." *Proceedings of the 8th International Conference on Multimodal Interfaces*. New York: Association for Computing Machinery, 2006.
Cohn, Jeffrey F., and Takeo Kanade. "Use of Automated Facial Image Analysis for Measurement of Emotion Expression." In *The Handbook of Emotion Elicitation and Assessment*, edited by James A. Coan and John J. B. Allen, 222–238. New York: Oxford University Press, 2007.
Cohn, Jeffrey F., Takeo Kanade, Tsuyoshi Moriyama, Zara Ambadar, Jing Xiao, Jiang Gao, and Hiroki Imamura. "Final Report, CIA Contract #2000-A128400-000: A Comparative Study of Alternative FACS Coding Algorithms" (November 9, 2001). http://mplab.ucsd.edu/grants/project1/publications/pdfs/CMUFinalReport.pdf.
Cole, Simon A. *Suspect Identities: A History of Fingerprinting and Criminal Identification*. Cambridge: Harvard University Press, 2001.
Cole, Simon A., and Henry N. Pontell. "'Don't Be a Low Hanging Fruit': Identity Theft as Moral Panic." In *Surveillance and Security: Technological Politics and Power in Everyday Life*, edited by Torin Monahan, 125–148. New York: Routledge, 2006.
Coté, Mark, and Jennifer Pybus. "Learning to Immaterial Labour 2.0: MySpace and Social Networks." *Ephemera* 7, no. 1 (2007): 88–106.
Crary, Jonathan. *Techniques of the Observer: On Vision and Modernity in the Nineteenth Century*. Cambridge: MIT Press, 1992.
Curry, Michael R. "The Profiler's Question and the Treacherous Traveler: Narratives of Belonging in Commercial Aviation." *Surveillance and Society* 1, no. 4 (2004): 475–499.
Czitrom, Daniel J. *Media and the American Mind: From Morse to McLuhan*. Chapel Hill: University of North Carolina Press, 1982.
Dalby, Simon. "Contesting an Essential Concept: Reading the Dilemmas in Contemporary Security Discourse." In *Critical Security Studies*, edited by Keith Krause and Michael C. Williams, 3–32. Minneapolis: University of Minnesota Press, 1997.
Dandeker, Christopher. *Surveillance, Power, and Modernity: Bureaucracy and Discipline from 1700 to the Present Day*. Cambridge: Polity Press, 1990.
Daston, Lorraine, and Peter Galison. "The Image of Objectivity." *Representations* 40 (1992): 81–128.

Datta, Ritendra, Dhiraj Joshi, Jia Li, and James Z. Wang. "Image Retrieval: Ideas, Influences, and Trends of the New Age." *ACM Computing Surveys* 40, no. 2 (2008): 1–60.

Dávila, Arlene. *Barrio Dreams: Puerto Ricans, Latinos, and the Neoliberal City*. Berkeley: University of California Press, 2004.

Davis, Mike. *City of Quartz*. New York: Vintage, 1992.

De Landa, Manuel. *War in the Age of Intelligent Machines*. New York: Zone Books, 1991.

Deleuze, Gilles. *Cinema 1: The Movement-Image*, translated by Hugh Tomlinson and Barbara Habberjam. Minneapolis: University of Minnesota Press, 1986.

Deleuze, Gilles, and Félix Guattari. *A Thousand Plateaus: Capitalism and Schizophrenia*, translated by Brian Massumi. Minneapolis: University of Minnesota Press, 1987.

Der Derian, James. *Virtuous War: Mapping the Military-Industrial-Media-Entertainment Network*. Boulder: Westview Press, 2001.

Donath, Judith. "Mediated Faces." In *Proceedings of the 4th International Conference on Cognitive Technology: Instruments of the Mind*, edited by Meurig Beynon, Chrystopher L. Nehaniv, and Kerstin Dautenhahn, 373–390. London: Springer-Verlag, 2001.

Douglas, Susan. *Inventing American Broadcasting, 1899–1922*. Baltimore: Johns Hopkins University Press, 1987.

Doyle, Aaron. "An Alternative Current in Surveillance and Control: Broadcasting Surveillance Footage of Crimes." In *The New Politics of Surveillance and Visibility*, edited by Kevin D. Haggerty and Richard V. Ericson, 199–224. Toronto: University of Toronto Press, 2006.

Dreyfus, Hubert. *What Computers Still Can't Do*. Cambridge: MIT Press, 1992.

Dror, Otniel E. "The Scientific Image of Emotion: Experience and Technologies of Inscription." *Configurations* 7, no. 3 (1999): 355–401.

Dubbeld, Lynsey. "The Role of Technology in Shaping CCTV Surveillance Practices." *Information, Communication, and Society* 8, no. 1 (2005): 84–100.

Edwards, Paul N. *The Closed World: Computers and the Politics of Discourse in Cold War America*. Cambridge: MIT Press, 1996.

Ekman, Paul. "Strong Evidence for Universals in Facial Expressions: A Reply to Russell's Mistaken Critique." *Psychological Bulletin* 115, no. 2 (1994): 268–287.

———. *Telling Lies: Clues to Deceit in the Marketplace, Politics, and Marriage*. New York: Norton, 2001.

Ekman, Paul, and Wallace Friesen. "Measuring Facial Movement with the Facial Action Coding System." In *Emotion in the Human Face*, edited by Paul Ekman, 178–211. Cambridge: Cambridge University Press, 1982.

Elias, Norbert. *The Civilizing Process*, translated by Edmund Jephcott. Rev. ed. Oxford: Blackwell, 1994.

Elmer, Greg. *Profiling Machines: Mapping the Personal Information Economy*. Cambridge: MIT Press, 2004.

Engle, Karen. "The Face of a Terrorist." *Cultural Studies <=> Critical Methodologies* 7, no. 4 (2007): 397–424.

Ericson, Richard V., and Kevin D. Haggerty. *Policing the Risk Society*. Toronto: University of Toronto Press, 1997.

Evans, David. *Paying with Plastic: The Digital Revolution in Buying and Borrowing*. Cambridge: MIT Press, 1999.

Farah, Martha J., Kevin D. Wilson, Maxwell Drain, and James N. Tanaka. "What Is 'Special' About Face Perception?" *Psychological Review* 105, no. 3 (1998): 482–498.
Fasel, Beat, and Juergen Luettin. "Automatic Facial Expression Analysis: A Survey." *Pattern Recognition* 36, no. 1 (2003): 259–275.
Foucault, Michel. *Discipline and Punish: The Birth of the Prison*, translated by Alan Sheridan. New York: Vintage, 1977.
———. *The History of Sexuality: An Introduction*, translated by Robert Hurley. New York: Vintage, 1978.
———. *Security, Territory, Population: Lectures at the College de France, 1977–78*, edited by Michel Senellart, translated by Graham Burchell. London: Palgrave, 2007.
———. *"Society Must Be Defended": Lectures at the Collège de France, 1975–1976*, translated by David Macey. New York: Picador, 2003.
———. "Technologies of the Self." In *Ethics: Subjectivity and Truth*, edited by Paul Rabinow, translated by Robert Hurley, 223–252. New York: New Press, 1997.
Fridlund, Alan. *Human Facial Expression: An Evolutionary View*. San Diego: Academic Press, 1994.
*Frye v. United States* (1923) 54 App. D. C. 46, 293 F. 1013.
Gandy, Oscar. *The Panoptic Sort: A Political Economy of Personal Information*. Boulder: Westview, 1993.
Garland, David. *The Culture of Control: Crime and Social Order in Contemporary Society*. Chicago: University of Chicago Press, 2001.
———. "The Limits of the Sovereign State: Strategies of Crime Control in Contemporary Society." *British Journal of Criminology* 36, no. 4 (1996): 445–471.
Giddens, Anthony. *The Constitution of Society: Outline of the Theory of Structuration*. Berkeley: University of California Press, 1984.
———. *Modernity and Self-Identity*. Palo Alto: Stanford University Press, 1991.
Gill, Pat. "Technostalgia: Making the Future Past Perfect." *Camera Obscura: A Journal of Feminism, Culture, and Media Studies* 14, nos. 1–2 (1997): 163–179.
Gilman, Sander. *The Face of Madness: Hugh W. Diamond and the Origin of Psychiatric Photography*. New York: Brunner/Mazel, 1976.
Gilroy, Paul. *"There Ain't No Black in the Union Jack": The Cultural Politics of Race and Nation*. London: Hutchinson, 1987.
Goffman, Erving. *Interaction Ritual: Essays in Face-to-Face Behavior*. New Brunswick, NJ: Aldine Transaction, 1967/2005.
Goldstein, A. J., L. D. Harmon, and A. B. Lesk. "Man-Machine Interaction in Human-Face Identification." *Bell System Technical Journal* 51, no. 2 (1972): 399–427.
Golumbia, David. *The Cultural Logic of Computation*. Cambridge: Harvard University Press, 2009.
Graham, Stephen. "The Software-Sorted City: Rethinking the 'Digital Divide.'" In *The Cybercities Reader*, edited by Stephen Graham, 324–333. New York: Routledge, 2004.
———. "Software-Sorted Geographies." *Progress in Human Geography* 29, no. 5 (2005): 562–580.
———. "Towards of Fifth Utility? On the Extension and Normalisation of Public CCTV." In *Surveillance, Closed Circuit Television, and Social Control*, edited by Clive Norris, Jade Moran, and Gary Armstrong, 89–112. London: Ashgate, 1998.

Graham, Stephen, and David Wood. "Digitizing Surveillance: Categorization, Space, Inequality." *Critical Social Policy* 23, no. 2 (2003): 227–248.
Gray, Mitchell. "Urban Surveillance and Panopticism: Will We Recognize the Facial Recognition Society?" *Surveillance and Society* 1, no. 3 (2003): 314–330.
Groombridge, Nic. "Stars of CCTV? How the Home Office Wasted Millions—A Radical 'Treasury/Audit Commission' View." *Surveillance and Society* 5, no. 1 (2008): 73–80.
Gunning, Tom. "In Your Face: Physiognomy, Photography, and the Gnostic Mission of Early Film." *Modernism/Modernity* 4, no. 1 (1997): 1–29.
Hacking, Ian. "How Should We Do the History of Statistics?" in *The Foucault Effect: Studies in Governmentality*, edited by Graham Burchell, Colin Gordon, and Peter Miller, 181–196. Chicago: University of Chicago Press, 1991.
———. "Making Up People." In *Reconstructing Individualism: Autonomy, Individuality, and the Self in Western Thought*, edited by Thomas C. Heller, Morton Sosna, and David E. Wellbery, 222–236. Palo Alto: Stanford University Press, 1986.
Haggerty, Kevin D., and Richard V. Ericson. "The New Politics of Surveillance and Visibility." In *The New Politics of Surveillance and Visibility*, edited by Kevin D. Haggerty and Richard V. Ericson, 3–25. Toronto: University of Toronto Press, 2006.
———. "The Surveillance Assemblage." *British Journal of Sociology* 51, no. 4 (2000): 605–622.
Haggerty, Kevin D., and Amber Gazso. "Seeing Beyond the Ruins: Surveillance as a Response to Terrorist Threats." *Canadian Journal of Sociology* 30, no. 2 (2005): 169–187.
Hall, Stuart. "Cultural Identity and Diaspora." In *Identity: Community, Culture, Difference*, edited by Jonathan Rutherford, 222–237. London: Lawrence and Wishart, 1990.
Hall, Stuart, Chas Critcher, Tony Jefferson, John Clarke, and Brian Roberts. *Policing the Crisis: Mugging, the State, and Law and Order*. London: Palgrave, 1978.
Hamilton, Peter, and Roger Hargreaves. *The Beautiful and the Damned: The Creation of Identity in Nineteenth-Century Photography*. London: Lund Humphries, 2001.
Hammonds, Evelyn. "New Technologies of Race." In *Processed Lives: Gender and Technology in Everyday Life*, edited by Jennifer Terry and Melodie Calvert, 74–85. New York: Routledge, 1997.
Hansen, Flemming, and Sverre Riis Christensen. *Emotions, Advertising, and Consumer Choice*. Copenhagen, Denmark: Copenhagen Business School Press, 2007.
Haraway, Donna. *Modest_Witness@Second_Millennium.FemaleMan©_Meets_OncoMouse™*. New York: Routledge, 1997.
———. "The Persistence of Vision." In *The Visual Culture Reader*, edited by Nicholas Mirzoeff, 677–684. New York: Routledge, 1998.
Harvey, David. *A Brief History of Neoliberalism*. New York: Oxford University Press, 2005.
———. *The Condition of Postmodernity*. Malden, MA: Blackwell, 1990.
———. "Flexible Accumulation Through Urbanization: Reflections on 'Post-modernism' in the American City." In *Post-Fordism: A Reader*, edited by Ash Amin, 361–386. Cambridge, MA: Blackwell, 1994.
Haxby, James V., Elizabeth A. Hoffman, and M. Ida Gobbini. "Human Neural Systems for Face Recognition and Social Communication." *Biological Psychiatry* 51, no. 1 (2002): 59–67.
Hay, James. "Designing Homes to Be the First Line of Defense: Safe Households, Mobilization, and the New Mobile Privatization." *Cultural Studies* 20, nos. 4–5 (2006): 349–377.

Hayles, N. Katherine. *How We Became Posthuman: Virtual Bodies in Cybernetics, Literature, and Informatics*. Chicago: University of Chicago Press, 1999.

Hempel, Leon, and Eric Töpfer. "The Surveillance Consensus: Reviewing the Politics of CCTV in Three European Countries." *European Journal of Criminology* 6, no. 2 (2009): 157–177.

Hill, Robert. "Retina Identification." In *Biometrics: Personal Identification in a Networked Society*, edited by Anil K. Jain, Ruud Bolle, and Sharath Pankanti, 123–141. Boston: Kluwer Academic, 1999.

Hobsbawm, Eric. *The Age of Extremes: A History of the World, 1914–1991*. New York: Vintage, 1994.

Hunter, Richard. *World Without Secrets: Business, Crime, and Privacy in the Age of Ubiquitous Computing*. New York: Gartner, 2002.

Iacono, William C. "Forensic 'Lie Detection': Procedures Without Scientific Basis." *Journal of Forensic Psychology Practice* 1, no. 1 (2001): 75–86.

Introna, Lucas D. "Disclosive Ethics and Information Technology: Disclosing Facial Recognition Systems." *Ethics and Information Technology* 7, no. 2 (2005): 75–86.

Introna, Lucas D., and Helen Nissenbaum. *Facial Recognition Technology: A Survey of Policy and Implementation Issues*. Center for Catastrophe Preparedness and Response, New York University, nd. http://www.nyu.edu/projects/nissenbaum/papers/facial_recognition_report.pdf.

Introna, Lucas D., and David Wood. "Picturing Algorithmic Surveillance: The Politics of Facial Recognition Systems." *Surveillance and Society* 2, nos. 2–3 (2004): 177–198.

Jarret, Kylie. "Interactivity Is Evil! A Critical Investigation of Web 2.0." *First Monday* 13, no. 3 (2008). http://firstmonday.org/htbin/cgiwrap/bin/ojs/index.php/fm/article/view/2140/1947.

Jay, Martin. *Downcast Eyes: On the Denigration of Vision in Twentieth-Century French Thought*. Berkeley: University of California Press, 1993.

———. "Scopic Regimes of Modernity." In *Vision and Visuality*, edited by Hall Foster, 3–23. Seattle: Bay Press, 1988.

Johnston, John. "Machinic Vision." *Critical Inquiry* 26, no. 1 (1999): 27–48.

Kanade, Takeo. *Computer Recognition of Human Faces*. Basel: Birkhauser, 1977.

Keen, Sam. *Faces of the Enemy: Reflections of the Hostile Imagination*. San Francisco: Harper & Row, 1986.

Keltner, Dacher, Terrie E. Moffit, and Magda Stouthamer-Leober. "Facial Expressions of Emotion and Psychopathology in Adolescent Boys." In *What the Face Reveals: Basic and Applied Studies of Spontaneous Expression Using the Facial Action Coding System (FACS)*, edited by Paul Ekman and Erika Rosenberg, 434–446. New York: Oxford University Press, 1997.

Kemp, Sandra. *Future Face: Image, Identity, Innovation*. London: Profile Books, 2004.

Klein, Naomi. *The Shock Doctrine: The Rise of Disaster Capitalism*. New York: Metropolitan, 2007.

Koskela, Hille. "Webcams, TV Shows, and Mobile Phones: Empowering Exhibitionism." *Surveillance and Society* 2, nos. 2–3 (2004): 199–215.

Krause, Keith, and Michael C. Williams, eds. *Critical Security Studies*. Minneapolis: University of Minnesota Press, 1997.

Latour, Bruno. *Aramis, or the Love of Technology*, translated by Catherine Porter. Cambridge: Harvard University Press, 1996.

———. “Mixing Humans and Non-humans Together: The Sociology of a Door-Closer.” *Social Problems* 35, no. 3 (1988): 298–310.
———. *Science in Action: How to Follow Scientists and Engineers Through Society*. Cambridge: Harvard University Press, 1987.
Lauer, Josh. “From Rumor to Written Record: Credit Reporting and the Invention of Financial Identity in Nineteenth-Century America.” *Technology and Culture* 49, no. 2 (2008): 301–324.
Levinas, Emmanuel. *Totality and Infinity: An Essay on Exteriority*, translated by Alphonso Lingis. Pittsburgh: Duquesne University Press, 1969.
Li, Stan Z., and Anil K. Jain. “Chapter 1: Introduction.” In *Handbook of Face Recognition*, edited by Stan Z. Li and Anil K. Jain, 1–11. New York: Springer, 2005.
Lisetti, Christine L., and Diane J. Schiano. “Automatic Facial Expression Interpretation.” *Pragmatics and Cognition* 8, no. 1 (2000): 185–235.
Littlefield, Melissa. “Constructing the Organ of Deceit: The Rhetoric of fMRI and Brain Fingerprinting in Post-9/11 America.” *Science, Technology, and Human Values* 34, no. 3 (2009): 365–392.
Lloyd, Martin, *The Passport: The History of Man's Most Traveled Document*. Thrupp Stroud, UK: Sutton, 2003.
Lyon, David. *Surveillance After September 11*. Cambridge: Polity, 2003.
———. *Surveillance Society: Monitoring Everyday Life*. Buckingham: Open University Press, 2001.
MacKenzie, Donald. *Inventing Accuracy: A Historical Sociology of Nuclear Missile Guidance*. Cambridge: MIT Press, 1993.
Magnet, Shoshana. “Bio-benefits: Technologies of Criminalization, Biometrics, and the Welfare System.” In *Surveillance: Power, Problems, and Politics*, edited by Sean Hier and Joshua Greenberg, 169–183. Vancouver: University of British Columbia Press, 2009.
Manning, Peter. “Information Technology and the Police.” In *Crime and Justice* 15, *Modern Policing*, edited by Michael H. Tonry and Norval Morris. Chicago: University of Chicago Press, 1992, 349–398.
Manovich, Lev. *The Language of New Media*. Cambridge: MIT Press, 2001.
Martin, Randy. *Financialization of Daily Life*. Philadelphia: Temple University Press, 2002.
Massey, Doreen. *Space, Place, and Gender*. Minneapolis: University of Minnesota Press, 1994.
Mattelart, Armand. *The Information Society*. Thousand Oaks, CA: Sage, 2003.
———. *Theories of Communication*. Thousand Oaks, CA: Sage, 1998.
Maxwell, Richard. “Surveillance: Work, Myth, and Policy.” *Social Text 83* 23, no. 2 (2005): 1–19.
McCahill, Michael. “Beyond Foucault: Towards a Contemporary Theory of Surveillance.” In *Surveillance, Closed Circuit Television, and Social Control*, edited by Clive Norris, Jade Moran, and Gary Armstrong, 41–68. Aldershot, UK: Ashgate, 1998.
———. *The Surveillance Web: The Rise of Visual Surveillance in an English City*. Portland: Willan, 2002.
McCarthy, Anna. “Closed Circuit Television” (Museum of Broadcast Communications, 2010). http://www.museum.tv/archives/etv/C/htmlC/closedcircui/closedcircui.htm.
McCormack, D. “Can Corporate America Secure Our Nation? An Analysis of the Identix Framework for the Regulation and Use of Facial Recognition Technology.” *Boston University Journal of Science and Technology Law* 9, no. 1 (2003): 128–155.

Metz, Christian. *The Imaginary Signifier: Psychoanalysis and the Cinema*, translated by Celia Britton et al. Bloomington: Indiana University Press, 1982.
Miller, Toby. *Makeover Nation: The Unites States of Reinvention*. Columbus: Ohio State University Press, 2008.
Mitchell, William J. *The Reconfigured Eye: Visual Truth in the Post-photographic Era*. Cambridge: MIT Press, 2001.
Monahan, Torin. "The Surveillance Curriculum: Risk Management and Social Control in the Neoliberal School." In *Surveillance and Security: Technological Politics and Power in Everyday Life*, edited by Torin Monahan, 109–124. New York: Routledge, 2006.
Mormino, Gary R., and George E. Pozzetta. *The Immigrant World of Ybor City: Italians and Their Latin Neighbors in Tampa, 1885–1985*. Urbana: University of Illinois Press, 1987.
Mosco, Vincent. *The Digital Sublime: Myth, Power, and Cyberspace*. Cambridge: MIT, 2004.
Muller, Benjamin J. "(Dis)qualified Bodies: Securitization, Citizenship and 'Identity Management.'" *Citizenship Studies* 8, no. 3 (2004): 279–294.
Murray, Janet. *Hamlet on the Holodeck: The Future of Narrative in Cyberspace*. Cambridge: MIT Press, 1997.
Nadesan, Majia. *Biopower, Governmentality, and Everyday Life*. New York: Routledge, 2008.
Nakamura, Lisa. "The Socioalgorithmics of Race: Sorting It Out in Jihad Worlds." In *The New Media of Surveillance*, edited by Shoshana Magnet and Kelly Gates, 147–161. New York: Routledge, 2009.
Negroponte, Nicholas. *Being Digital*. New York: Vintage, 1995.
Nissenbaum, Helen, and Lucas D. Introna. "Shaping the Web: Why the Politics of Search Engines Matters." *Information Society* 16, no. 3 (2000): 169–185.
Norberg, Arthur L., and Judy E. O'Neill. *Transforming Computer Technology: Information Processing for the Pentagon, 1962–1986*. Baltimore: Johns Hopkins University Press, 1996.
Norris, Clive, and Gary Armstrong. *The Maximum Surveillance Society*. Oxford: Berg, 1999.
Norris, Clive, Michael McCahill, and David Wood. "The Growth of CCTV: A Global Perspective on the International Diffusion of Video Surveillance in Publicly Accessible Space." *Surveillance and Society* 2, no. 2 (2004): 110–135.
Norris, Clive, Jade Moran, and Gary Armstrong. "Algorithmic Surveillance: The Future of Automated Visual Surveillance." In *Surveillance, CCTV, and Social Control*, edited by Clive Norris, Jade Moran, and Gary Armstrong, 255–275. Aldershot, UK: Ashgate, 1998.
O'Harrow, Robert, Jr. *No Place to Hide: Behind the Scenes of Our Emerging Surveillance Society*. New York: Free Press, 2005.
Ong, Aihwa. *Neoliberalism as Exception: Mutations in Citizenship and Sovereignty*. Durham: Duke University Press, 2006.
O'Shaughnessy, John, and Nicholas Jackson O'Shaughnessy. *The Marketing Power of Emotion*. Oxford: Oxford University Press, 2003.
Peters, John Durham. *Speaking into the Air: A History of the Idea of Communication*. Chicago: University of Chicago Press, 1999.
Phillips, P. Jonathan, Patrick J. Raus, and Sandor Z. Der. *FERET (Face Recognition Technology) Recognition Algorithm Development and Test Results* (Army Research Laboratory, October 1996). ARL-TR-995, http://www.nist.gov/humanid/feret/doc/army_feret3.pdf.
Picard, Rosalind. *Affective Computing*. Cambridge: MIT Press, 1998.
Pine, B. Joseph, II. *Mass Customization: The New Frontier in Business Competition*. Cambridge: Harvard Business School Press, 1992.

Poster, Mark. "Databases as Discourse, or Electronic Interpellations." In *Computers, Surveillance, and Privacy*, edited by David Lyon and Elia Zureik, 175–192. Minneapolis: University of Minnesota Press, 1996.

Raab, Charles D. "Joined-Up Surveillance: The Challenge to Privacy." In *The Intensification of Surveillance: Crime, Terrorism, and Warfare in the Information Age*, edited by Kirstie Ball and Frank Webster, 42–61. London: Pluto Press, 2003.

Robertson, Craig. "A Documentary Regime of Verification: The Emergence of the U.S. Passport and the Archival Problematization of Identity." *Cultural Studies* 23, no. 3 (2009): 329–354.

Robins, Kevin. *Into the Image: Culture and Politics in the Field of Vision*. London: Routledge, 1996.

Robins, Kevin, and Frank Webster. *Times of the Technoculture: From the Information Society to the Virtual Life*. New York: Routledge, 1999.

Rose, Nikolas. *Inventing Our Selves: Psychology, Power, and Personhood*. New York: Cambridge University Press, 1998.

———. *Powers of Freedom: Reframing Political Thought*. New York: Cambridge University Press, 1999.

Rosenberg, Erika L. "Introduction: The Study of Spontaneous Facial Expressions in Psychology." In *What the Face Reveals: Basic and Applied Studies of Spontaneous Expression Using the Facial Action Coding System*, edited by Paul Ekman and Erika Rosenberg, 3–17. New York: Oxford University Press, 1997.

Rule, James. *Private Lives and Public Surveillance*. London: Allen Lane, 1973.

Rushton, Richard. "Response to Mark B. N. Hansen's Affect as Medium, or the 'Digital-Facial-Image.'" *Journal of Visual Culture* 3, no. 3 (2004): 353–357.

———. "What Can a Face Do? On Deleuze and Faces." *Cultural Critique* 51 (2002): 219–237.

Rygiel, Kim. "Protecting and Proving Identity: The Biopolitics of Waging War Through Citizenship in the Post-9/11 Era." In *(En)gendering the War on Terror: War Stories and Camouflaged Politics*, edited by Krista Hunt and Kim Rygiel, 146–176. Aldershot, UK: Ashgate, 2006.

Sakai, T., M. Nagoa, and S. Fukibayashi. "Line Extraction and Pattern Recognition in a Photograph." *Pattern Recognition* 1, no. 3 (1969): 233–248.

Salter, Mark B. "When the Exception Becomes the Rule: Borders, Sovereignty, and Citizenship." *Citizenship Studies* 12, no. 4 (2008): 365–380.

Schiller, Dan. *Digital Capitalism: Networking the Global Market System*. Cambridge: MIT Press, 2000.

———. *How to Think About Information*. Urbana: University of Illinois Press, 2007.

Schiller, Herbert I. *Mass Communications and American Empire*. 2nd ed. Boulder: Westview Press, 1992.

Scott, James. *Seeing Like a State: How Certain Schemes to Improve the Human Condition Have Failed*. New Haven: Yale University Press, 1998.

Seagal, Debra. "Tales from the Cutting-Room Floor: The Reality of Reality-Based Television." *Harper's Magazine* (November 1993): 50–57.

Sekula, Allan. "The Body and the Archive." *October* 39 (Winter 1986): 3–64.

———. "Photography Between Labour and Capital." In *Mining Photographs and Other Pictures, 1948–1968: A Selection from the Negative Archives of Shedden Studio, Glace Bay,*

*Cape Breton*, edited by Benjamin H. D. Buchloh and Robert Wilkie, 193–268. Halifax: Press of the Nova Scotia College of Art and Design, 1983.

Simon, Jonathan. *Governing Through Crime: How the War on Crime Transformed American Democracy and Created a Culture of Fear*. Cambridge: Oxford University Press, 2007.

Simpson, Timothy. "Communication, Conflict, and Community in an Urban Industrial Ruin." *Communication Research* 22, no. 6 (1995): 700–719.

Sobieszek, Robert A. *A Ghost in the Shell: Photography and the Human Soul*. Cambridge: MIT Press, 1999.

Sontag, Susan. *On Photography*. New York: Picador, 1977.

Sorkin, Michael. "Introduction: Variations on a Theme Park." In *Variations on a Theme Park: The New American City and the End of Public Space*, edited by Michael Sorkin, xi–xv. New York: Hill and Wang, 1992.

Stanley, Jay, and Barry Steinhardt. *Drawing a Blank: The Failure of Facial Recognition Technology in Tampa, Florida* (American Civil Liberties Union, January 3, 2002). https://www.aclu.org/FilesPDFs/drawing_blank.pdf.

Sterne, Jonathan. *The Audible Past: Cultural Origins of Sound Reproduction*. Durham: Duke University Press, 2003.

———. "Bourdieu, Technique, and Technology." *Cultural Studies* 17, no. 3 (2003): 367–389.

———. "Enemy Voice." *Social Text 96* 26, no. 3 (2008): 79–100.

Suwa, Motoi, Noboru Sugie, and Keisuke Fujimura. "A Preliminary Note on Pattern Recognition of Human Emotional Expression." In *Proceedings of the 4th International Joint Conference on Pattern Recognition, November 7–10, 1978, Kyoto, Japan* (New York: Institute of Electrical and Electronics Engineers, 1979): 408–410.

Tagg, John. *The Burden of Representation: Essays on Photographies and Histories*. Minneapolis: University of Minnesota Press, 1988.

Telotte, J. P. *A Distant Technology: Science Fiction Film and the Machine Age*. Hanover: Wesleyan University Press, 1999.

Terranova, Tiziana. "Free Labor: Producing Culture for the Digital Economy." *Social Text 63* 18, no. 2 (2003): 33-57.

Thobani, Sunera. *Exalted Subjects: Studies in the Making of Race and Nation*. Toronto: University of Toronto Press, 2007.

Thrift, Nigel. *Non-representational Theory: Space, Politics, Affect*. New York: Routledge, 2008.

Tian, Ying-Li, Takeo Kanade, and Jeffrey Cohn. "Facial Expression Analysis." In *Handbook of Face Recognition*, edited by Stan Z. Li and Anil K. Jain, 247–275. New York: Springer, 2004.

Torpey, John. *The Invention of the Passport: Surveillance, Citizenship, and the State*. Cambridge: Cambridge University Press, 2000.

Turk, Matthew. "Computer Vision in the Interface." *Communications of the ACM* 47, no. 1 (January 2004): 61–67.

Turk, Matthew, and Alex Pentland. "Eigenfaces for Recognition." *Journal of Cognitive Neuroscience* 3, no. 1 (1991): 71–86.

Turkle, Sherry. *Life on the Screen: Identity in the Age of the Internet*. New York: Touchstone, 1995.

Turow, Joseph. *Breaking Up America: Advertisers and the New Media World*. Chicago: University of Chicago Press, 1997.

———. *Niche Envy: Marketing Discrimination in the Digital Age*. Cambridge: MIT Press, 2006.
Tytler, Graeme. *Physiognomy in the European Novel: Faces and Fortunes*. Princeton: Princeton University Press, 1982.
Van der Ploeg, Irma. "Biometrics and Privacy: A Note on the Politics of Theorizing Technology." *Information, Communication, and Society* 6, no. 1 (2003): 85–104.
Van Oenen, Gijs. "A Machine That Would Go of Itself: Interpassivity and Its Impact on Political Life." *Theory and Event* 9, no. 2 (2006). http://muse.jhu.edu/journals/theory_and_event/v009/9.2vanoenen.html.
Virilio, Paul. *The Vision Machine*, translated by Julie Rose. Bloomington: Indiana University Press, 1994.
———. *War and Cinema: The Logistics of Perception*, translated by Patrick Camiller. London: Verso, 1989.
Wacquant, Loïc. "The Penalisation of Poverty and the Rise of Neo-liberalism." *European Journal on Criminal Policy and Research* 9, no. 4 (2001): 40
Weber, Max. *The Theory of Social and Economic Organization*, translated by A. M. Henderson and Talcott Parsons. Glencoe, IL: Free Press, 1947.
Webster, William. "CCTV Policy in the UK: Reconsidering the Evidence Base." *Surveillance and Society* 6, no. 1 (2009): 10–22.
Weizenbaum, Joseph. *Computer Power and Human Reason*. New York: Freeman, 1976.
Welchman, John. "Face(t)s: Notes on Faciality." *ArtForum* (November 1988): 131–138.
Whitson, Jennifer R., and Kevin D. Haggerty. "Identity Theft and the Care of the Virtual Self." *Economy and Society* 37, no. 4 (2008): 572–594.
Williams, Chris. "Police Surveillance and the Emergence of CCTV in the 1960s." *Crime Prevention and Community Safety* 5, no. 3 (2003): 27–37.
Wilson, Christopher. *Cop Knowledge: Police Power and Cultural Narrative in Twentieth-Century America*. Chicago: University of Chicago Press, 2000.
Winner, Langdon. "Who Will We Be in Cyberspace?" *Information Society* 12, no. 1 (1996): 63–72.
Winston, Brian. *Media Technology and Society: A History: From the Telegraph to the Internet*. New York: Routledge, 1998.
Wood, David Murakami. "A New 'Baroque Arsenal'? Surveillance in a Global Recession." *Surveillance and Society* 6, no. 1 (2009): 1–2.
Zhao, Wen-Yi, Rama Chellappa, P. Jonathan Phillips, and Azriel Rosenfeld. "Face Recognition: A Literature Survey." *ACM Computer Surveys* 35, no. 4 (2003): 399–458.
Žižek, Slavoj. *The Plague of Fantasies*. New York: Verso, 1997.
Zureik, Elia, and Karen Hindle. "Governance, Security, and Technology: The Case of Biometrics." *Studies in Political Economy* 73 (2004): 113–137.

# Index

# About the Author

KELLY A. GATES is Assistant Professor in the Department of Communication and the Science Studies Program at the University of California, San Diego. She is the coeditor of *The New Media of Surveillance*.